INTRODUCING EARTH SCIENCE

A practical approach to geology

INTRODUCING EARTH SCIENCE

A practical approach to geology

James Bradberry

Basil Blackwell

First published 1985

Published by
Basil Blackwell Limited
108 Cowley Road
Oxford OX4 1JF
England

Typeset in 10 on 12 point Trump
by Oxford Publishing Services, Oxford
Printed in Great Britain by Bell and Bain Ltd., Glasgow

ISBN 0 631 90600 2

CONTENTS

Acknowledgements	vi
Preface	1
1 An introduction to the Earth	2
2 Plate tectonics	9
3 The changing Earth	16
4 Mountain building	21
5 Earthquakes	25
6 Volcanoes	33
7 Minerals	42
8 Rocks	54
9 Soil	69
10 Fossils	76
11 The search for wealth	90
12 Seas and oceans	95
13 Limestone	100
14 Water movement	106
15 Britain: Past and present	113
16 Using rocks	124
17 Fuel resources	130
Appendix 1: Mineral identification (computer program)	140
Appendix 2: List of suppliers	141
Answers to exercises	142
Bibliography	145
Index	147

ACNOWLEDGEMENTS

The author and publishers wish to thank the following for permission to reproduce:

Fig 1.1 Palomar Observatory Photograph; Fig 1.2 John Shelton; Fig 2.6 Institute of Geological Sciences; Fig 4.2 John Shelton; Figs 5.2, 6.1 U.S. Geological Survey; Fig 6.2 Heather Angel/Biophotos; Fig 6.10 Johan Henrik Piepgrass/Iceland Tourist Board; Figs 6.13, 7.1, 7.2 Institute of Geological Sciences; Fig 7.11 Rida Photo Library; Figs 7.12, 7.13 Institute of Geological Sciences; Figs 8.1, 8.2 Jim Bradberry; Fig 8.3 Institute of Geological Sciences; Fig 8.7 Eric Kay; Fig 8.10 Jim Bradberry; Figs 8.11, 8.12, 8.13, 8.15 Institute of Geological Sciences; Fig 8.17a Con Gillen; Figs 8.18, 8.21, 8.24 Institute of Geological Sciences; Fig 8.25 U.S. Geological Survey; Figs 10.1, 10.2 British Museum (Natural History); Fig 10.3 Rida Photo Library; Fig 10.5 P.V. Glob/Forhistorisk Museum, Denmark; Fig 10.6 Institute of Geological Sciences; Fig 10.7 and extract Granada Publishing Ltd for material from *Lucy*, D. C. Johanson and M. A. Edey, text based on pp.19–22 and artwork based on the charts on pp.186–7 and 290–1; Figs 10.10a,b Rida Photo Library; Figs 10.11a,b British Museum (Natural History); Fig 10.12 Rida Photo Library; Fig 10.13 Imag; Fig 10.14 Institute of Geological Sciences; Fig 10.15 Rida Photo Library; Figs 10.16a, 10.17 Institute of Geological Sciences; Figs 10.18a,b Rida Photo Library; Fig 10.19 Institute of Geological Sciences; Fig 10.21 Leicestershire Museums; Figs 10.22 Smithsonian Institution, Washington; Fig 11.1 R.T.Z. Services Limited; Fig 11.8 Barnabys Picture Library; Fig 12.3 Institute of Oceanographic Sciences; Figs 13.2, 13.3 Institute of Geological Sciences; Fig 13.4 Sedgwick Museum, Cambridge; Fig 13.5 Institute of Geological Sciences; Fig 13.6 Jim Bradberry; Fig 13.9 Dales Photographs; Fig 13.10 Jim Bradberry; Fig 13.11 Institute of Geological Sciences; Fig 13.12 Foster Yeoman Limited; Fig 14.6 Institute of Geological Sciences; Fig 14.7 Jim Bradberry; Fig 14.8 Aerofilms; Fig 14.9 Lancaster Guardian Series; Fig 15.1 Jim Bradberry; Fig 15.3 Institute of Geological Sciences; Fig 15.6a Barnabys Picture Library; Fig 15.6b Institute of Geological Sciences; Fig 15.11 Bradford Washburn/Boston Museum of Science; Fig 15.12 Institute of Geological Sciences; Figs 15.13, 15.14 Aerofilms; Figs 15.15, 15.16, 15.17 Institute of Geological Sciences; Figs 15.18, 15.19, 15.20, 15.21, 15.22 Aerofilms; Fig 16.1 International Photobank; Fig 16.3 Ken Trathen; Fig 16.5 Barnabys Picture Library; Fig 17.1 Institute of Geological Sciences; Fig 17.5 Barnabys Picture Library; Fig 17.6 National Coal Board.
Cover Woodmansterne Publications Ltd.

PREFACE

This book is an introduction to the processes that have helped to form our planet. It has been written for students taking geology and geography examinations at 16+ and GCSE, and covers the requirements of the geology syllabuses at this level. Students of environmental science will also find much of use and interest in this book.

Students are likely to start studying geology with little previous knowledge of the subject and so particular care has been taken to make the text lively and readable, and to include up-to-date and interesting case-study material. Each chapter contains suggestions for practical activities which are simple to carry out and need little specialised equipment. The activities are particularly important in the chapters on rock and mineral identification where they provide opportunities for valuable geological investigation. At the end of every chapter there are graded exercises and word puzzles.

If the readers of this book develop a greater interest in, and awareness of, the Earth and its processes then the writing of it will have been worthwhile.

I am grateful for the advice and guidance received from many different sources, in the preparation of this book. In particular I would like to thank Adrian Brierley of the Geology Department, University of Manchester, and J. W. Elder of the Geophysics Department, University of Manchester, for their help on the sections relating to geothermal energy; and Paul Green of Fair Oak School, Rugeley, for permission to adapt the mineral computer program that he devised.

James Bradberry
1985

1 AN INTRODUCTION TO THE EARTH

From outer space, Neil Armstrong, the American astronaut, described the Earth as 'a beautiful jewel in space'. Most of what he saw was water: the deep blue of the oceans and the reflective white wisps of cloud. He also saw the brownish-green colours of the land masses and the brilliant white of the polar ice caps. The Earth's surface temperatures are such that most water stays in liquid form. If the earth was slightly cooler, much of the water would be stored in huge polar ice caps.

THE ATMOSPHERE

The atmosphere is a continuous layer of gases which surrounds the earth. It is made up as follows:

nitrogen 78%
oxygen 21%
other gases 1% – mainly carbon dioxide, water vapour, hydrogen.

The atmosphere exists because the Earth's gravitational pull prevents the gases escaping into space. It is important for the following reasons:

a) it filters out harmful radiation from the sun;
b) it acts as a store for the sun's heat;
c) storms and air movements spread heat and moisture across the Earth's surface, so maintaining even climatic conditions;
d) it protects the Earth from meteorite falls.

Fig 1.1 The Galaxy – this consists of millions of stars. Our sun is one of them.

THE SOLAR SYSTEM

The sun is a *star* in the Milky Way galaxy (Fig. 1.1). The *solar system* is the sun encircled by nine major planets, including Earth, and several thousand tiny planets called *asteroids*. These planets revolve around the sun in nearly circular paths called *orbits*. The sun is really quite a small star, compared with other stars – but it is huge compared with the planets. The sun contains 99.9% of the mass of the *solar system*. This means that it is millions of times bigger than the planet Earth. Even the largest planet, Jupiter, is just a speck by comparison with the sun.

The sun's heat and light is made by nuclear reactions in its interior. These reactions raise the temperature deep inside the sun to about 36 million degrees centigrade. The rate of the sun's radiation is so great that about three million tonnes of matter is converted into energy every second.

The inner planets are Mercury, Venus, Earth and Mars. They are the rocky planets containing heavier materials such as iron, magnesium, and silicon compounds.

The outer planets are Jupiter, Saturn, Uranus, Neptune and Pluto. Apart from Pluto they are larger than the inner planets and are made mainly of lighter materials such as hydrogen and helium. The outer planets are much colder than the inner ones because they are further from the sun.

There are many tiny pieces of rock called *meteors* in the solar system. These meteors are 'shooting stars' and glow as they burn up in the Earth's atmosphere. Sometimes larger meteors hit the Earth's surface – these are called *meteorites*.

METEORITES

Danger: Falling Rocks!

On 30 November 1954 Mrs Hewlett Hodges, of Alabama, USA, was sitting quietly on the settee after lunch. Suddenly there was a crashing sound. An object weighing about 8 kilograms came through the roof. It rebounded off a radio and struck her on the left thigh. Mrs Hodges was unlucky because although millions of meteors enter the Earth's atmosphere every day, few of them are actually noticed. Most meteors are very small and burn up long before they hit the ground. They can be seen by the bright trail they leave as they streak across the night sky. Even bigger meteors usually break up in the atmosphere or explode. Their remains eventually reach the ground as very fine dust. There is a constant fall-out of this space dust and several thousand tonnes filter down to the Earth annually.

Fig 1.2 The Meteor Crater, Arizona

Very occasionally a large meteor survives its passage through the atmosphere as Mrs Hodges found out.

How often do meteorites fall to Earth?

There are approximately 500 falls per year over the whole Earth. Many of these meteorite falls go unnoticed because they either happen at night, or they hit the Earth in remote uninhabited areas. Most of them fall into the sea. Only about 150 are likely to fall on the land and of these only about four have any chance of being found.

What they look like

The arrival of a meteorite is impressive and is often accompanied by light and sound effects. Even in daylight a bright fireball is visible, followed by a trail of gases and dust. This 'tail' is brightly luminous at night and is often tinged with green, red and yellow. The sound effects can be heard up to 80 km from the eventual impact site. People who have witnessed a meteorite fall have described loud thundery noises like rapid gun fire, an express train, whistlings and cracklings. These varied noises end with the thud of impact.

Meteor Crater, Arizona

In the middle of the Arizona Desert is a large crater (Fig 1.2). It must have been formed by the impact of a huge meteor, striking the Earth with incredible force. The crater is 1300 m across and over 180 m deep, with the raised rim 40 m high.

The meteorite that caused this crater exploded on impact, scattering into many millions of fragments. We know this because rock fragments have formed and fused together to form many lumps of silica glass. Small amounts of diamond have also been found; they formed under the tremendous heat and pressure of the impact. This happened about 10 000 years ago. Very large meteors with a mass greater than 100 tonnes can survive in one piece and strike the earth. Partial melting takes place as they pass through the atmosphere, due to friction. However, they retain most of their original speed, striking the Earth at over 150 000 miles per hour. Meteors weighing less than one tonne usually break up into smaller fragments. These either burn up because of friction or they are considerably slowed down. Therefore, small meteorites that reach the surface do not produce very deep craters. Most meteoric material reaching the Earth's surface is fine dust which does not form craters at all.

Where they come from

It is thought that meteorites represent material that was present when the solar system first came into being 5000 million years ago. Most of this original material combined to form the planets very early in time. In one sense, this combining process is still going on every time another meteorite is compelled by gravity to fall toward the Earth.

The effect of the Earth's atmosphere

Many of the planets are heavily cratered because of meteor bombardment. The craters on the moon are very common features and pockmark the surface. This is also true of Mercury and Mars and some of the outer planets. The recent American space probes have revealed similar cratering on the larger 'moons' of Jupiter and Saturn. Why is it that the Earth seems not to be so heavily cratered? The answer lies in the fact that the Earth has an atmosphere. This has two main effects.

Firstly, the envelope of gases shields the surface, causing objects to burn up due to friction. There is no such protection on a planet like the moon, where there is hardly any atmosphere.

Secondly, the winds, ice, rain and rivers are constantly *eroding* (wearing away) the Earth's surface rocks. This can only happen if the planet has an atmosphere. Because erosion is so effective at levelling irregularities such as crater walls and hollows, only relatively recent craters have survived.

AGE OF THE EARTH

The sheer vastness of geological time is difficult to grasp. Look carefully at Fig 1.3. You will see that the Earth has existed for about 4700 million years. Simple forms of life appeared about 3000 million years ago. To give a better sense of the scale of time, the main events are displayed to a scale of 12 months. The Earth's past history is divided into *geological eras*. These in turn are divided into *periods*. All the periods together represent approximately the last 1/10 of the Earth's time span. It is interesting to note that on the monthly time scale, the last glaciers disappeared from the British Isles at about 11.59 pm on 31 December. The Romans conquered Britain and left in about the last 13 seconds before the end of the year.

The beginning of the Cambrian period is marked by evidence of the quite sudden appearance of plant and animal life. Many animals and plants had begun to develop hard parts, which were preserved as *fossils* in the rock layers.

Compared to the vast age of the Earth, humans are insignificant. We appeared only in the last 1/100 of the Earth's total time span. On the monthly scale, what date in December might this have been?

HOW THE EARTH BEGAN

The sun with its surrounding planets came into being about 5000 million years ago, as a huge cold *nebula* (cloudy mass). This broke up to form a large central nebula with surrounding rotating 'clouds'. The gas and dust in each cloud began to condense and eventually formed the planets. The central solar nebula also contracted becoming much denser. Then it began to heat up and radiate light, and so the sun was formed.

Small amounts of new material were steadily added to the mass of each planet by collision. Large planets were able to hold lighter gases in their gravitational fields. Lighter gases in the nebula were dispersed by the radiating pressure of sunlight (solar wind).

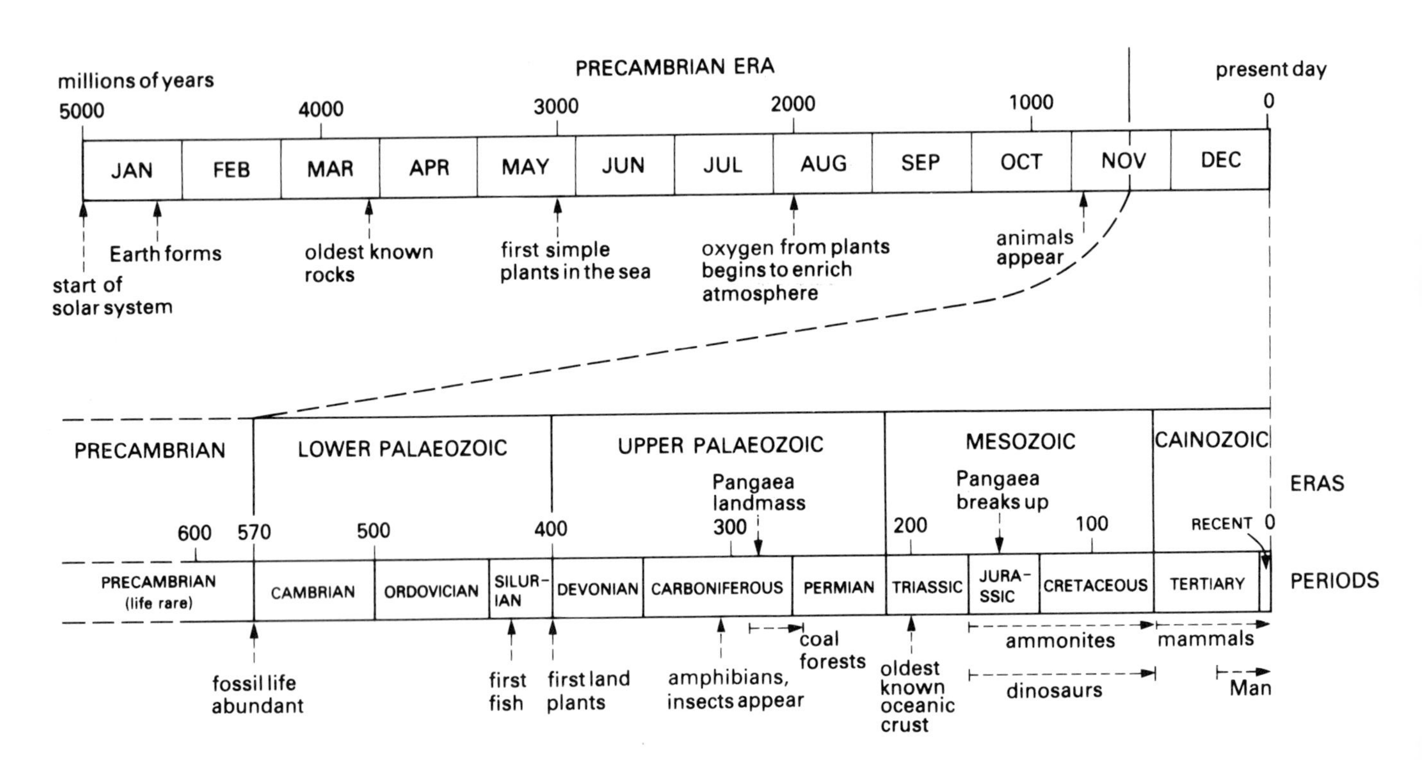

Fig 1.3 The age of the Earth

The early Earth was completely solid. The interior then began to heat up because of radioactive decay of minerals and the central core melted. Denser materials sank to form the nickel/iron *core*; lighter materials formed the rocky *mantle* and *crust*. The importance of these layers is described in the next section.

STRUCTURE OF THE EARTH (Fig 1.4)

Central core

The *central* core is made of very hot and dense material. The temperature is probably over 3000 °C yet it remains solid. This is because of the great pressure – three million times that at the Earth's surface. The core has a high metal content. It is a mixture of iron and nickel with sulphur. It has a high *density*: one cubic centimetre (cm^3) of the core would have a mass of at least 10 gram (g), that is, 10 g/cm^3. By comparison water has a density of 1 g/cm^3.

Outer core

This is a liquid layer. It is made of the same material as the central core but because there is less pressure it exists in a liquid state, although the temperature is about the same.

Magnetic field

The Earth has quite a strong magnetic field (Fig 1.5). This is caused by the circulation and flow of liquids in the outer core. The Earth's magnetism is much the same as that of a bar magnet.

The mantle

Surrounding the outer core, is a solid rock layer, 2881 km thick. Its average density is 4.5 g/cm^3. Rocks in the lower mantle are denser than upper-mantle rocks because they are compressed by a greater weight of rocks. Although the rocks are at very high temperatures, they are prevented from melting by the high pressure.

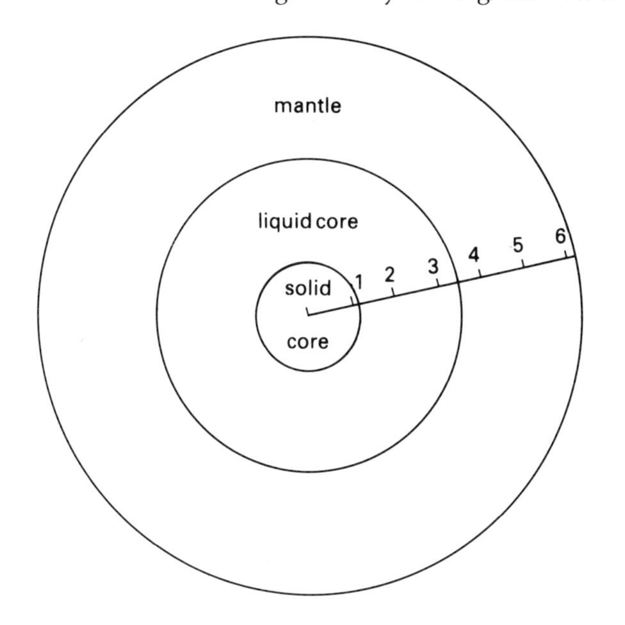

Fig 1.4 The structure of the Earth (linear scale is in 000 kms)

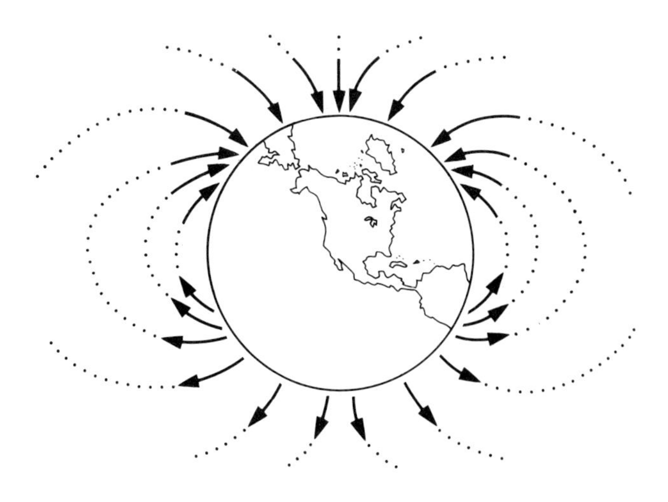

Fig 1.5 The Earth's magnetic field

Activity 1: Magnetism

Sprinkle iron filings on a sheet of paper placed above a bar magnet. Note how the filings align themselves in the magnetic field. Why do you think they behave the way they do?

Very slow movements can take place in the solid mantle, known as *convection currents*. Hot rocks rise very slowly towards the surface where they cool and eventually sink. See Activity 2.

The idea that solid material can flow may seem rather strange but think of the way plasticine or putty behaves. Movement of solid material in this way is known as *plastic flow*.

Rising convection currents in the mantle have two important effects:

a) they produce very hot rocks just below the surface. Here there is less pressure so the rocks melt, forming *magma*.

b) the sideways movement of convection currents in the mantle causes movements in the crust at the surface. The moving mantle pulls the crust along above it.

The crust

The crust is a very thin layer of surface rock. It is less dense and cooler than the mantle beneath. There are two distinct kinds of crust.

The thinner basaltic *oceanic crust* is found beneath the oceans. It is 8 km thick and covers 65% of the Earth's surface. The remaining 35% of the surface is covered by less dense *continental crust*, which rises to

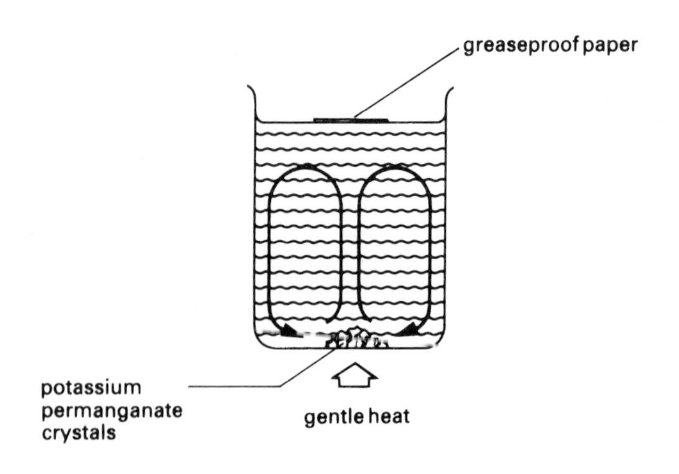

Fig 1.6 Demonstrating convection currents

Activity 2: Convection (Fig. 1.6)

Gently heat a beaker of water with a few crystals of potassium permanganate. The water at the base of the beaker expands as it is heated; it becomes less dense and rises towards the surface. The cold, denser water at the surface sinks. The convection currents can be seen as the potassium permanganate crystals slowly dissolve and stain the rising water.

1 Float a piece of greased paper in the middle. What happens to the floating paper? Can you explain why?
2 Make a sketch of this activity, and fully describe the process.
3 How is this activity a model of what is happening in the Earth?

form the land masses. It is 35 km thick on average (Fig 1.7).

The oceanic crust is much newer than continental crust. Nowhere is it older than 200 million years. The oldest continental crust is up to 19 times older (3800 million years is the oldest rock yet found). Oceanic basaltic crust forms directly from melted mantle material. Continental crust is made of many kinds of different rocks. These rocks have been contorted and folded by countless Earth movements over the years.

Shield areas

The oldest land areas are called *shields*, and represent the earliest 'continents'. They are made of once-molten rocks and early sediments, greatly altered by heat and pressure. Over millions of years there have been regular additions of continental material to the shield edges. The most recent additions are the folding and uplifting of sea-floor sediments to produce the long fold-mountain chains, eg, the Himalayas and Andes (see Chapter 4). These movements began about 50 million years ago.

The fold mountains and shield areas are worn away by forces of *erosion* such as wind, ice and rivers. The rock fragments are carried away as sand and mud, and spread out over lowlands and ocean basins. These sediments eventually form new rock layers. In time they may also become new fold mountains.

Vertical movements

The Earth's crust is in effect, floating on the denser mantle below. The thicker masses of continental crust behave rather like large icebergs – underneath each elevated section is a much thicker mass or 'root' supporting it. Each section of the Earth's crust finds its own buoyancy level. This will vary depending on its thickness. This kind of balance is called *isostasy*. It is easier to understand by looking at the behaviour of ice cubes of differing thickness floating in water (Fig 1.8).

If the top of the ice cube is removed, the cube bobs up, adjusting its height in the water. This kind of

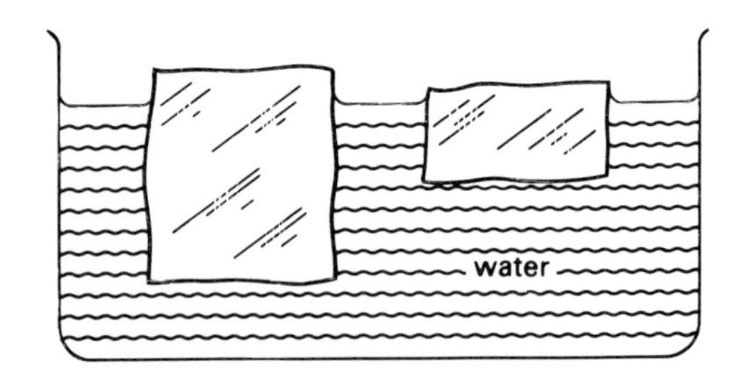

Fig 1.8 The ice cubes float with most of their bulk below water level. The thicker ice cube has a greater height above water level but extends more deeply into the water.

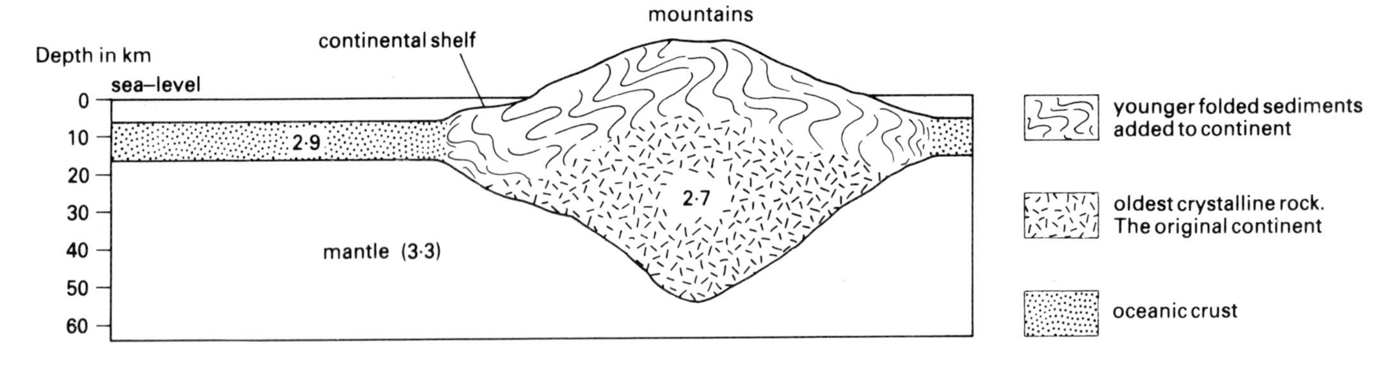

Fig 1.7 Continental and oceanic crust. The density is shown in g/cm³. The lighter crust effectively 'floats' on the mantle.

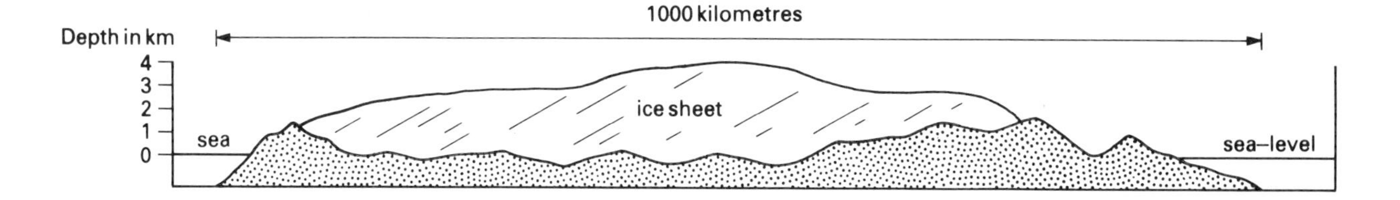

Fig 1.9 The Greenland Ice Sheet. The weight of the ice has caused the landmass to sink. If the ice was to suddenly melt, what would happen to the level of the Greenland Continent?

vertical movement happens if anything disturbs the isostatic balance. For instance, if the top of a mountain range is eroded, some of its mass is removed. That part of the crust will then slowly rise. If material is added to part of the crust, the extra mass will depress the crust further into the mantle (Fig 1.9).

Exercise 1: Inside the Earth

Make a copy of this crossword and then try to solve it. (*Solution on page 142*)

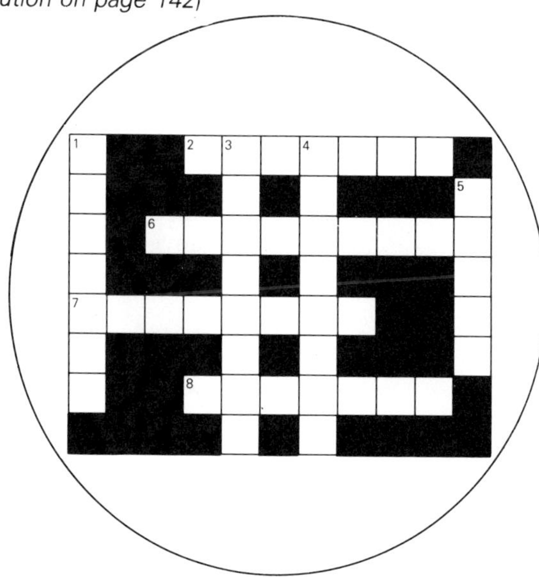

Across

2 The ________ less dense continental crust floats on the mantle (7)
6 The atmosphere shields the earth from meteors and harmful ________ (9)
7 This gas makes up 78% of the atmosphere (8)
8 They glow as they burn up in the atmosphere (7)

Down

1 Compared to continental crust, this kind of crust is much younger and thinner (7)
3 Some of this gas is present in the atmosphere and in the outer planets (8)
4 Planets with no atmosphere are often heavily ________ by meteorite falls (8)
5 The four ________ planets are composed of rocky materials (5)

Exercise 2

Write out and complete the following statements using the words listed below.

1 The sun is just one of the millions of ________ in the ________.
2 About 500 ________ fall to earth each year.
3 The earth formed ________ million years ago.
4 Fossils are rare in the ________ period.
5 The first land plants appeared at the start of the ________ period.
6 The Jurassic and Cretaceous periods were the Age of the ________ and ________.
7 A nebula is a cloud of ________ and dust.
8 Material in the nebula began to ________ together to form the planets.
9 To begin with the Earth was ________ then it began to heat up and melt.
10 The central ________ is made of iron and nickel. It has a high ________.
11 The ________ core is the only complete liquid layer inside the earth.
12 The Earth's ________ field is caused by movements of fluid in the outer core.
13 Convection currents take place in the ________.
14 The ________ areas of the continents are made of very old rocks, much altered by heat and ________.
15 If the rocks of a landmass are eroded the continent will ________ because of isostatic balance.

ammonites pressure outer density core gas stars 4700 galaxy cling Devonian Precambrian meteorites mantle magnetic rise dinosaurs shield solid

Exercise 3

1 Explain fully the nature and importance of the Earth's atmosphere.
2 a) What are meteors and meteorites?
b) Why is a study of meteorites of value in giving clues as to how the solar system began?
3 Copy the circles (Fig 1.10) and use the information in Fig 1.3 to draw two earth-time clocks.
a) Use your copy of clock A to record the main events throughout time.
b) Clock B is scaled to record the last 600 million years in 12 hours (1 hour = 50 million years). Divide the circle

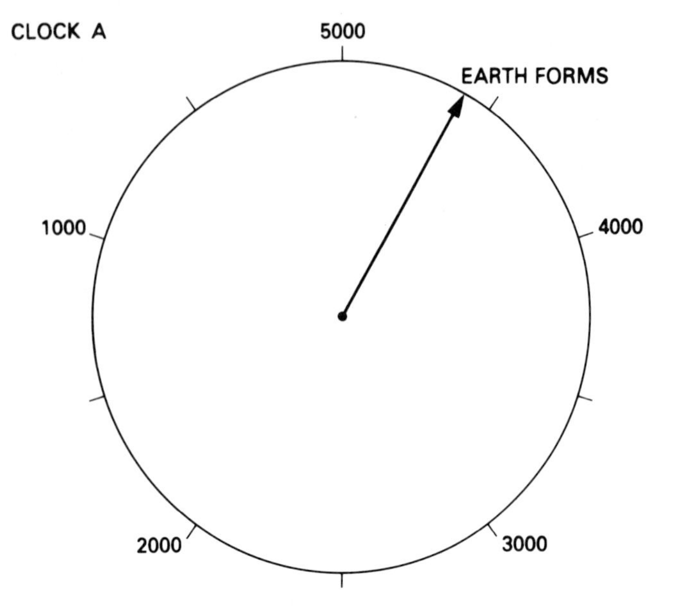

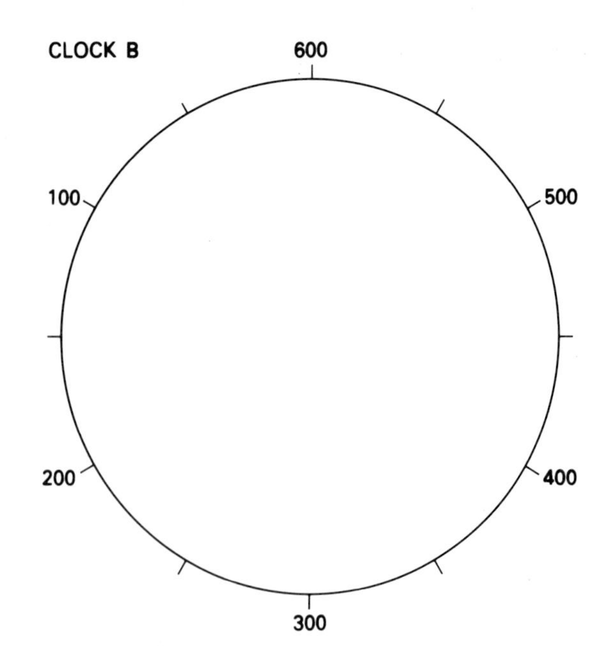

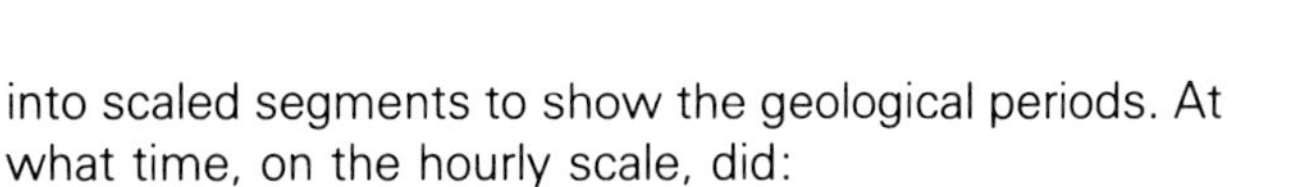

into scaled segments to show the geological periods. At what time, on the hourly scale, did:

(i) fish appear
(ii) land plants appear
(iii) insects appear?

c) For what length of time did dinosaurs exist?

d) How long have (i) mammals (ii) human beings, been on earth?

e) How long ago did the oldest oceanic crust form?

4 Briefly describe the formation of the Earth and the solar system.
5 Copy Fig 1.4 and briefly describe each layer of the Earth.
6 What causes the Earth's magnetic field? Make a sketch of the pattern produced by iron filings over a magnet. Compare your sketch to Fig 1.5 of the Earth's magnetic field, is it the same?
7 Why does hot material rise toward the surface in a convection current?
8 What is the most likely source of heat for the mantle convection currents?
9 Summarise the main differences between continental and oceanic crust in a table under the headings *Continental* and *Oceanic*.
10 What is meant by a 'shield' area?
11 Explain what is meant by isostatic balance. Use diagrams to illustrate your answer.

2 PLATE TECTONICS

There is a lot of evidence to support the idea that the Earth's surface is divided into a number of segments called *plates*. There are seven major world plates, plus a number of smaller ones. Each plate is a rigid, moving slab of rock. It is made of oceanic or continental crust, with a rigid layer of upper mantle welded on to its underside. The plates move slowly in response to convection currents in the plastic mantle below (Fig 2.1). As they move, they jostle each other for position. This means that central regions of plates remain quite stable but many changes take place at plate edges. These changes are known as *tectonic processes*. They include the folding and buckling of rocks under intense sideways pressure; the slipping of rocks along cracks called *faults*; the shuddering movements that cause *earthquakes*; and the melting of rock to produce *volcanoes*. Look carefully at the world distribution of earthquakes (Fig 2.2). Compare it to Fig 2.3. The linear pattern of earthquakes clearly shows where the plate boundaries are.

The way that all the plates move can be illustrated by removing the peel in segments from a whole orange. If the loose peel is then put back in position, and one segment is pushed sideways across the surface, it will cause every other segment on the face of the orange to move. The significance of this is that a plate on one side of the earth may cause related movements on the other side of the globe.

The effect of convection currents

Hot material slowly rises from the lower mantle to the surface, spreads out sideways and eventually cools and sinks (Fig 2.1). Upward convection pulls the plates apart at site A. Hot mantle material in the form of liquid magma continuously wells up from below. As the magma cools, it becomes part of the plates. At site B, the oceanic plate sinks below the thicker continental mass, to become part of the mantle again. Lighter layers of sediment on top of the denser oceanic plate, tend to buckle and fold upwards – they are added to the continental landmass nearby.

Rates of movement

Plates are moving at an average rate of six cm/year. This is about the rate at which your fingernails are growing. It may not sound very fast but in one million years this amounts to 60 km of movement! One million years is a very short time in the Earth's total history. How much movement could there be in the average human lifespan of 70 years?

Plate movement

It is possible to recognise three main kinds of movement (Fig 2.4).

a) Plates moving apart, as magma wells up from the

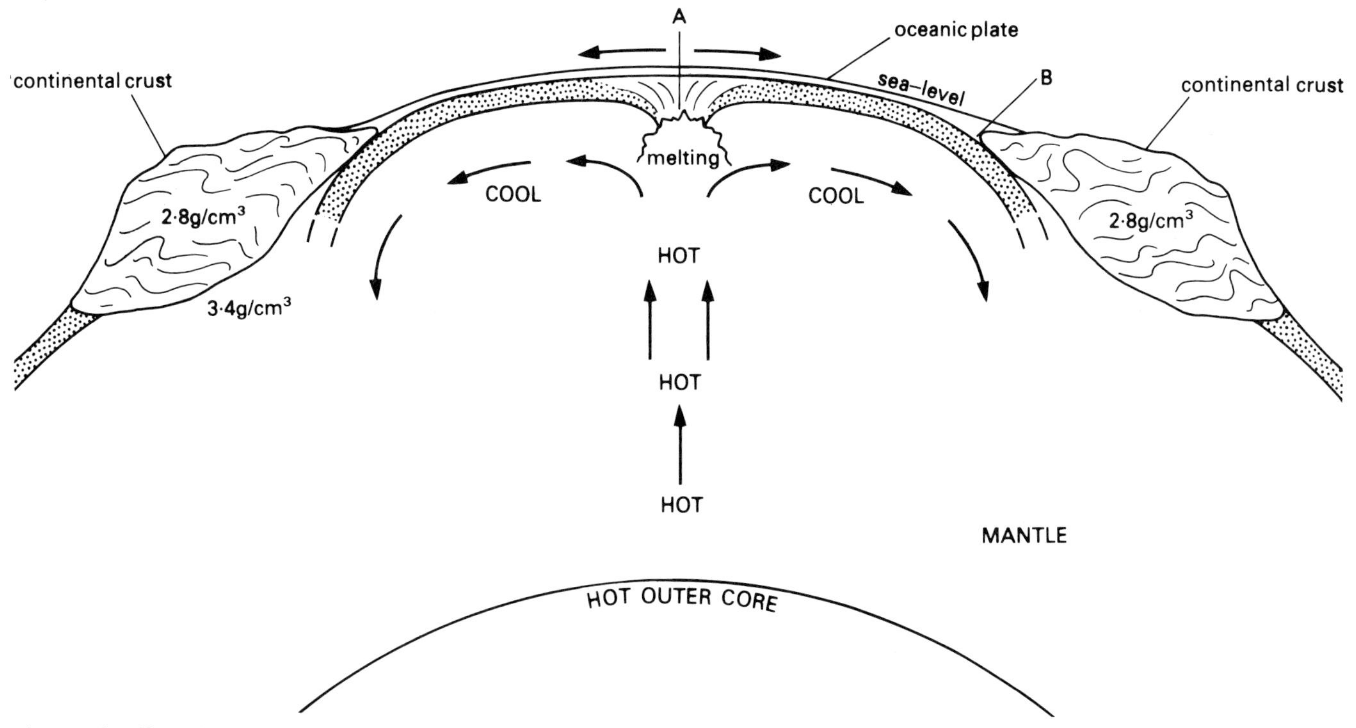

Fig 2.1 The effect of convection currents

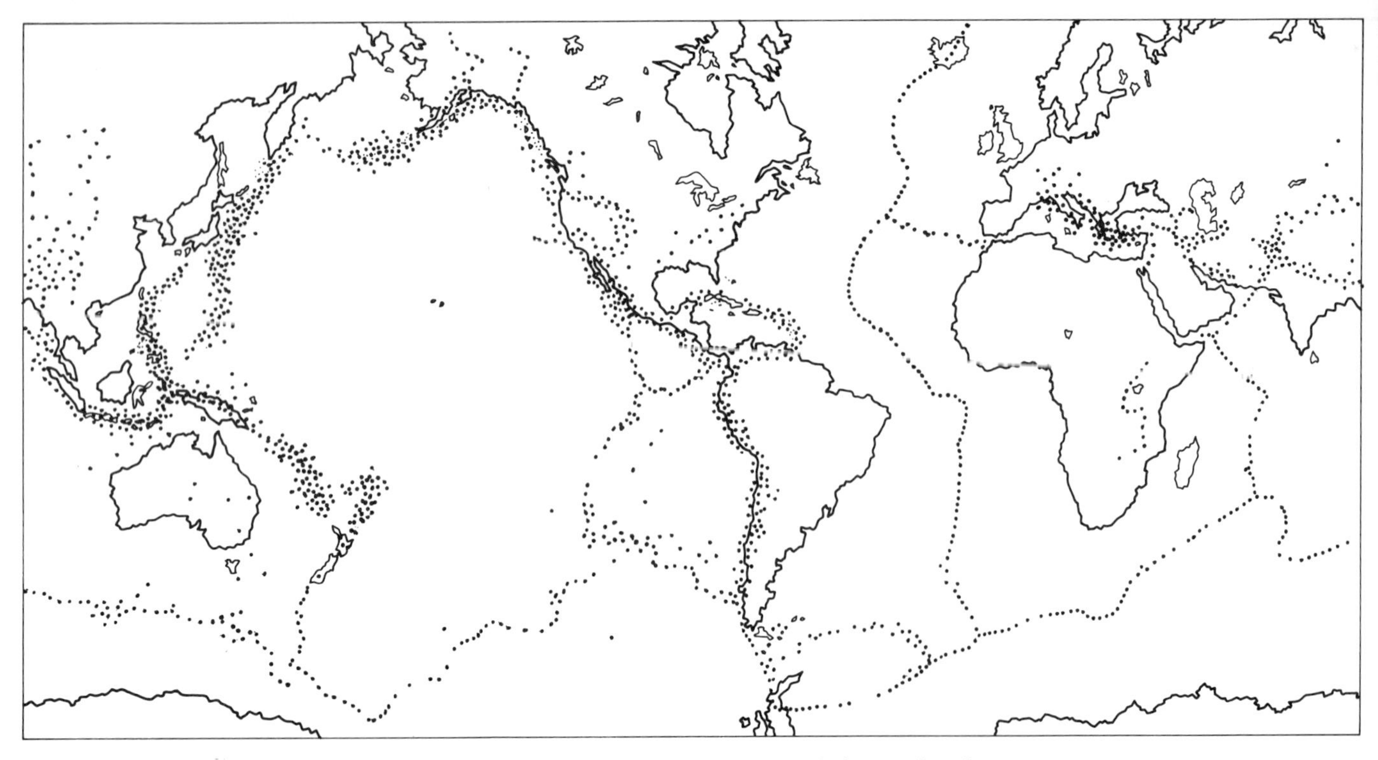

Fig 2.2 The world distribution of earthquakes. Each dot represents the site (epicentre) of an earthquake.

mantle to form new oceanic crust. These are *constructive* plate margins. Surface area is added to the Earth's surface.

b) Plates moving together – the collision causes one plate to slide beneath the other. The rocks are buckled, and surface area is lost. These are *destructive* plate margins.

c) Plates slide past each other along huge sideways-slip faults called *transform* faults. Surface area is neither lost nor gained. These are *conservative* plate margins.

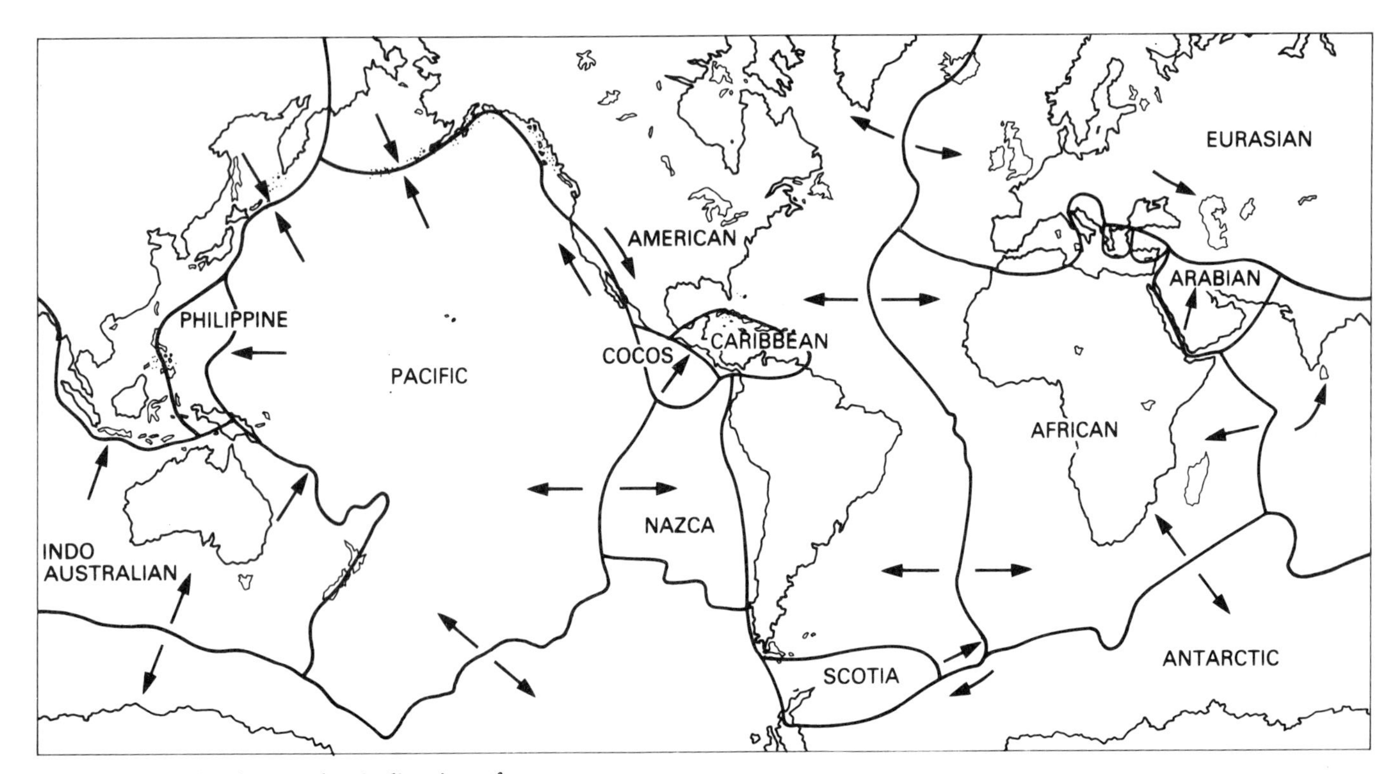

Fig 2.3 The Earth's plates, and main directions of movement

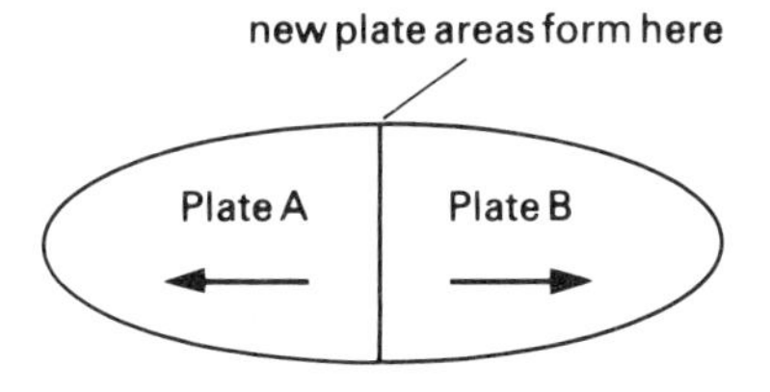

(a) Plates moving apart (constructive margin). The rocks are stretched by tensional forces.

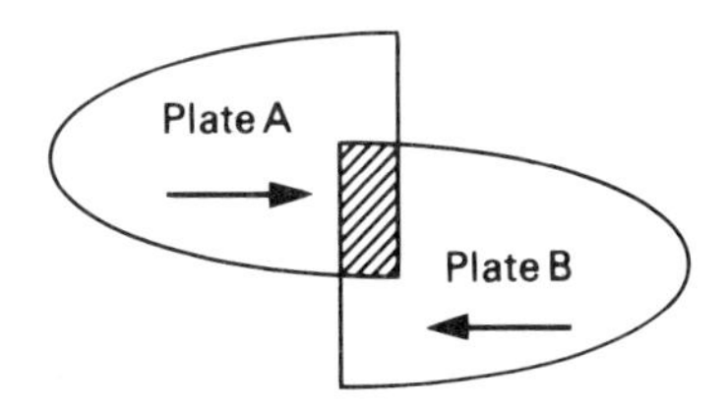

(b) Plates in collision (destructive margin). One plate slides beneath the other. The rocks are buckled by compression.

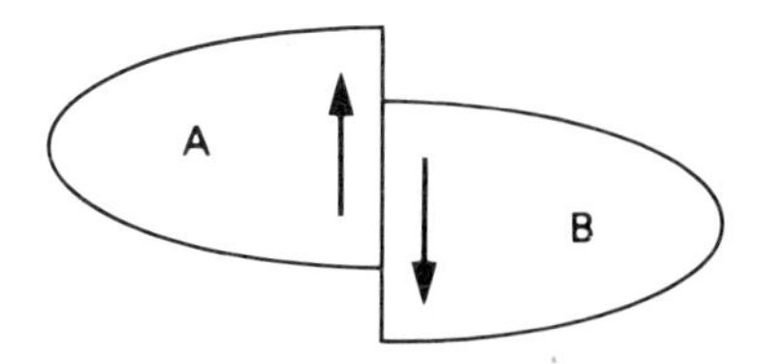

(c) Plates sliding past each other (conservative margin). The plates build up stress along transform faults. Sudden movements release this pent–up energy, causing earthquakes.

Fig 2.4 Different kinds of plate movement

Plates moving apart

The tension caused by plates moving away from each other causes the crust to thin, and lessens the pressure on the hot mantle material beneath. This allows the mantle rocks to melt to produce *magma,* which then cools as it reaches the surface to form the new oceanic crust. The zone in which new rock is formed is called the *spreading centre.* The addition of new rock to the plates and their movement away from the spreading centre, is known as *sea-floor spreading.*

Ocean ridges

The heated and expanded rocks at a spreading centre form a *linear ridge* on the ocean floor. In the Atlantic Ocean this ridge is found in a central position halfway between America and Africa. Ocean ridges form the longest and largest uplifted features on the Earth's surface. They tend to go unnoticed because it is rare for them to rise above sea level. The crest of an ocean ridge is marked by lines of parallel, near vertical cracks called *faults* and a *rift valley* (Fig 2.5). The faults are the result of stretching and tension. As the plates are pulled apart, wedges of the Earth's crust slide downwards. This helps to take up the tension. In fact the Earth's crust actually *increases* in surface area by this kind of faulting. In addition to the faulting there is intense and frequent volcanic activity as magma is injected into cracks and fault lines.

Most volcanic activity is *submarine.* A huge quan-

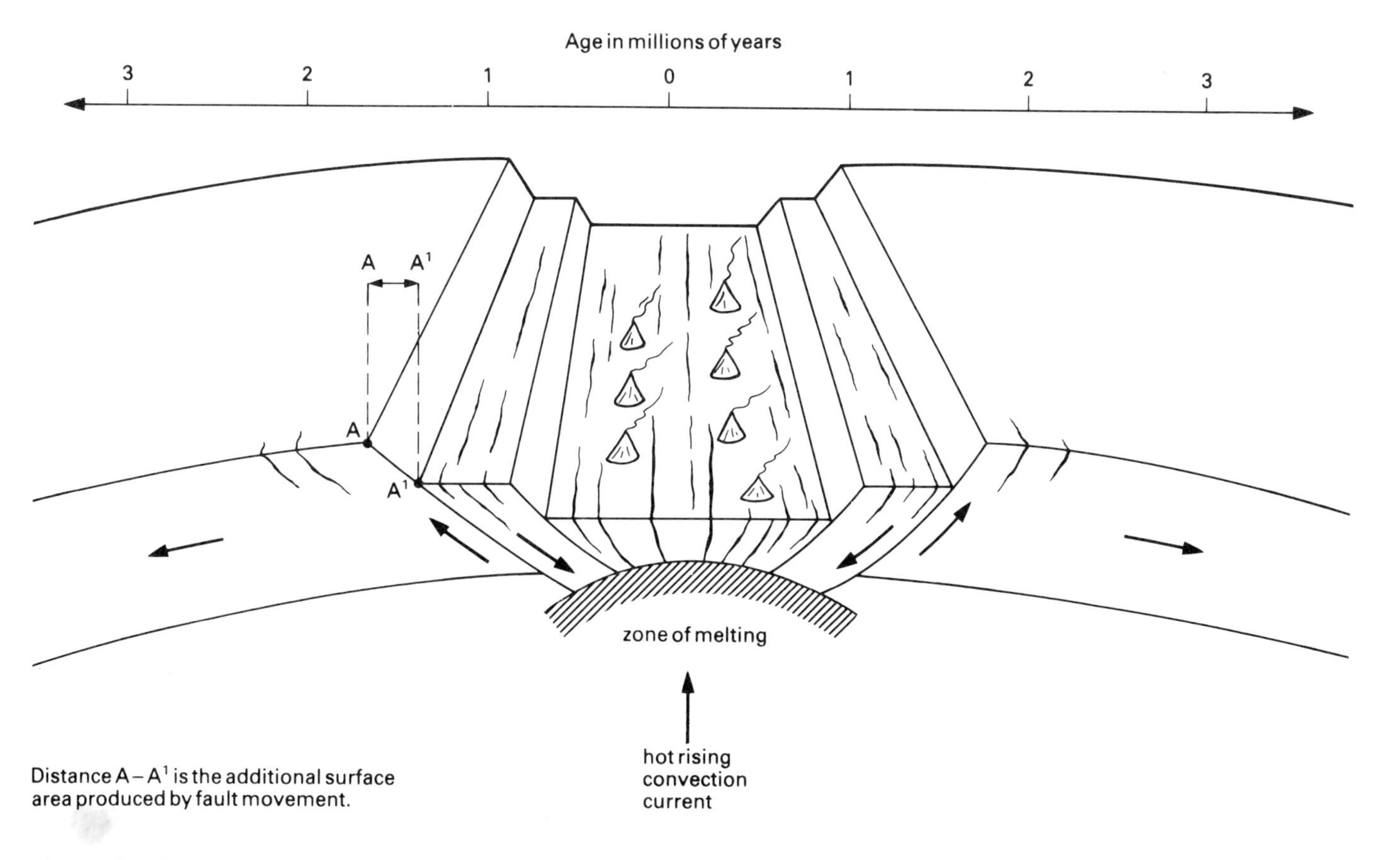

Fig 2.5 The rift valley along the crest of an ocean ridge. The expanded rocks cause uplift and produce a ridge. Tensional faulting and magma injection add surface area to the plates as they move apart. In Iceland this feature can be seen above sea level – see line A–B, Fig 2.8

Fig 2.6 Pillow lavas

tity of molten rock known as *lava* is poured out onto the sea floor. It then instantly hardens by contact with cold sea water to form successive *pillow* structures (Fig 2.6). Each 'bubble' of lava forms a hard outer crust but pressure of molten lava breaks through the 'bubble' to form a new structure which again instantly hardens. This kind of behaviour can be demonstrated using a burning candle (Fig 2.7).

The unusual case of Iceland

The island of Iceland sits on top of the mid-Atlantic ridge system. It is one of the few places on the Earth where an oceanic ridge rises above sea level (Fig 2.8). The tensional faults have produced a rift valley running through the centre of Iceland. This central valley is very active and there are many outpourings of basaltic magma. The magma wells up from long *fissures* that run parallel to the central rift valley. All the rocks in this area have formed within the last 20 000 years. The rocks outside the valley are progressively older to the east and west. Surveys have shown that Iceland is widening at a rate of two cm/year.

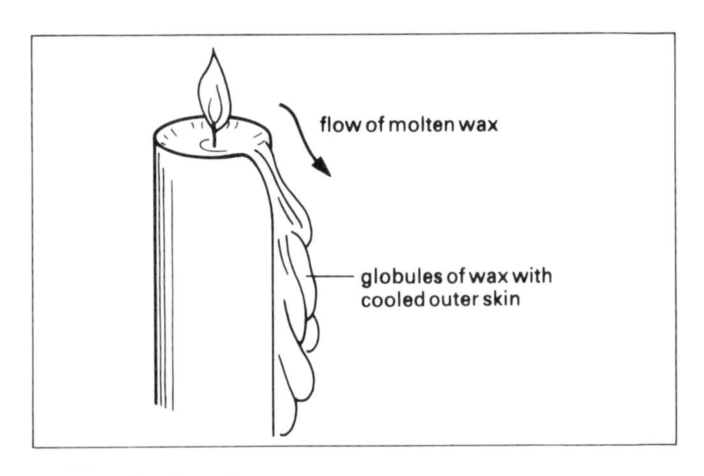

Fig 2.7 'Pillow lavas' in wax

Magnetic stripes and sea-floor spreading

Studies of the ocean floor rocks to the south of Iceland show a pattern of *magnetic stripes* on either side of the spreading ridge. As the new molten rock wells up from the mantle, it cools and goes solid. The small iron crystals within the molten rock line up with the direction of the Earth's magnetic field at the time of cooling. At intervals during the Earth's history the magnetic field has completely reversed itself – the south and north magnetic poles changing position. These reversals are recorded as symmetrical *stripes* on either side of the spreading ridge (Fig 2.9). Magnetic stripes, and the discovery that the youngest rocks are at the ridge crest, are important pieces of evidence. They support the idea that plates move apart as the newly-formed sea floor spreads sideways.

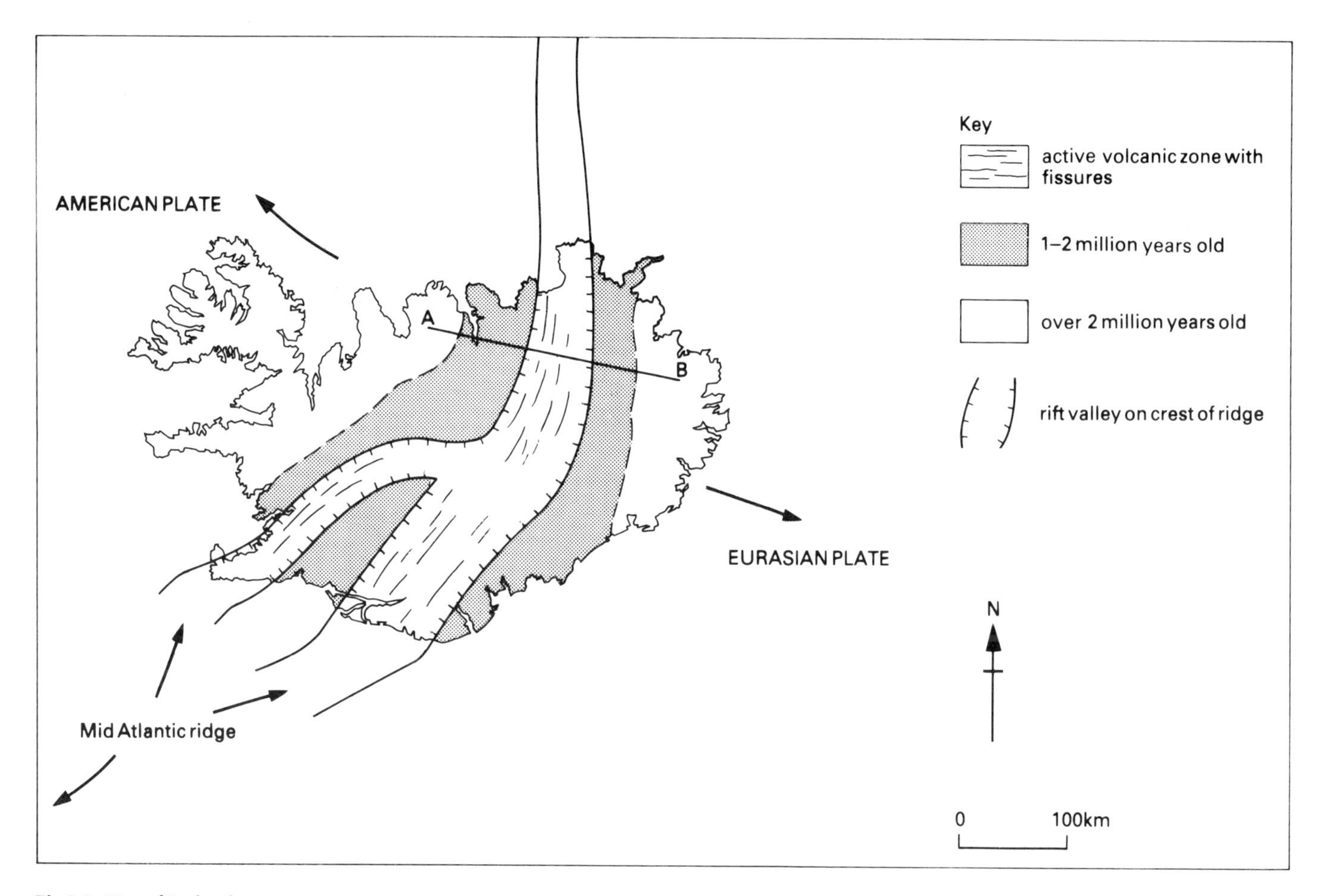

Fig 2.8 Map of Iceland

Plates in collision

Whenever an oceanic plate collides with the edge of a thicker or stronger plate, it sinks beneath it (Fig 2.10). This process is called *subduction*. The oceanic plate is dense and thin. It readily descends back into the mantle, where it came from in the first place. Subduction produces a deep *trench* on the ocean floor. The Marianas Trench in the Pacific is over 11 km deep. It is produced by the subduction of the Pacific plate beneath the Phillipine plate. In this case both plates are oceanic. There are also examples of oceanic plate subducting beneath continental plates. Can you see any on the map (Fig 2.3)? However, because of their low density, continental plate areas can never be lost by subduction. Why is this? (See Figs 1.7 and 2.1)

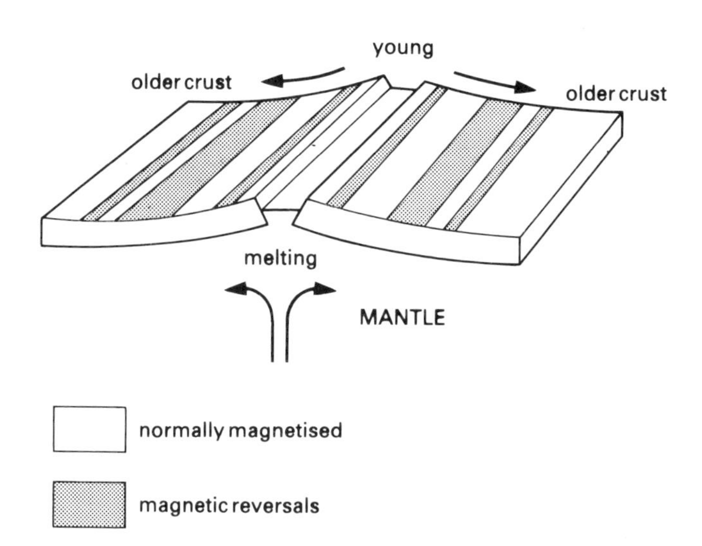

Fig 2.9 The symmetrical magnetic stripes of an ocean ridge system

The position of a subducting plate is clearly marked by regular earthquakes. These are caused by the shuddering movements of the plate as it descends. A zone of partial melting is produced between the top surface of the descending oceanic plate and the overriding plate. Melting occurs due to frictional heat, and to water contained in the oceanic basalts, acting as a *flux*. In other words, the water lowers the melting point. Pockets of magma are produced, known as *plutons*. Plutons are less dense than the rocks surrounding them, and so they rise. They cut their way through the overlying crust. When they reach the surface, chains of volcanic *island arcs* are produced. These are parallel to the trench on the landward side. Japan and the Aleutian islands are examples of island arcs.

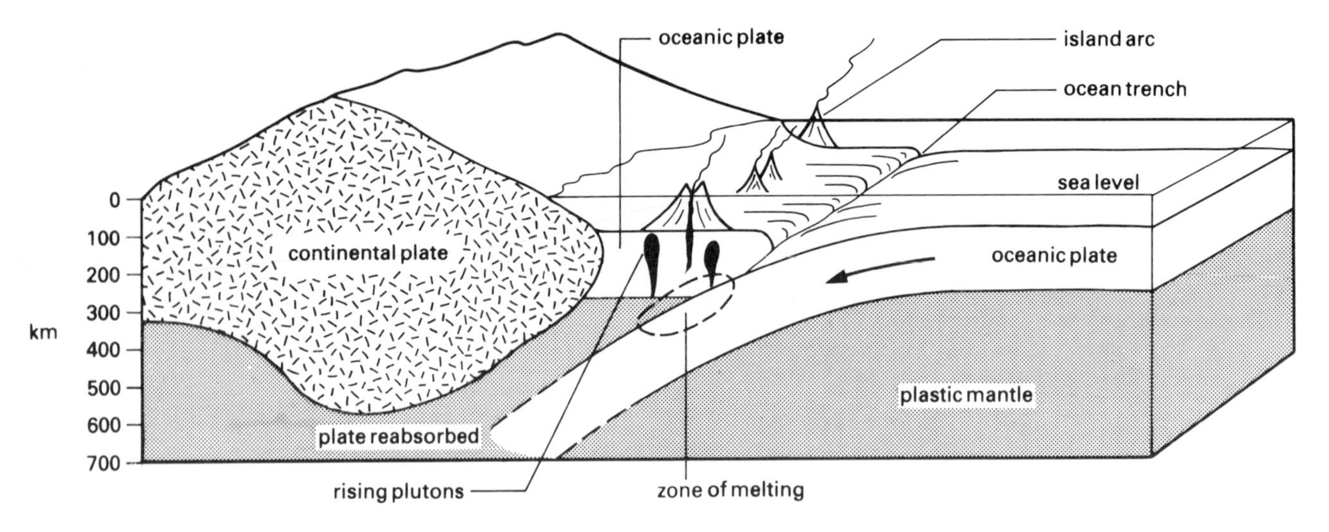

Fig 2.10 Oceanic plate collides with oceanic plate and returns to the mantle. The process of subduction produces earthquakes, melting of rocks, plutons, volcanic island arcs, and deep ocean trenches.

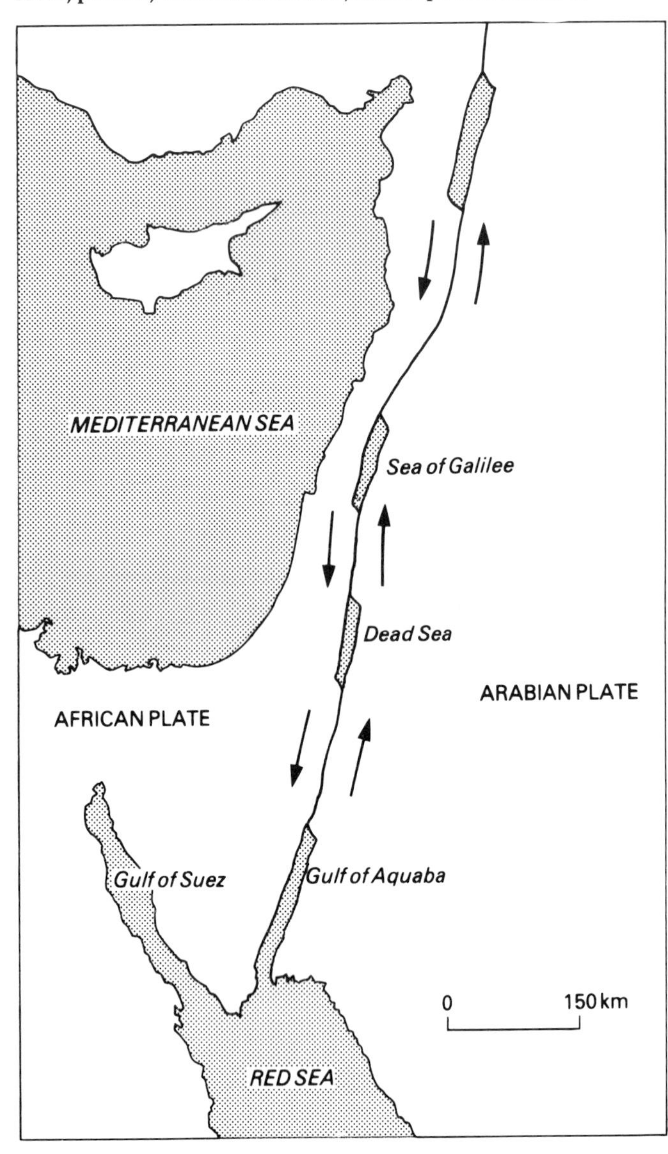

Fig 2.11 The Arabian plate is sliding northwards past the African plate along a huge transform fault. This has opened up a number of gulfs and inland seas.

Plates sliding past each other

There are many places on the Earth where plates are sliding past each other along huge sideways-slip faults called *transform* faults (Fig 2.11). Most of these faults cut across the ocean floor but sometimes they are seen on land. The plates on either side of the fault are moving in opposing directions. The sudden jerky movements cause severe earthquakes, eg, the San Francisco earthquake of 1906 on the San Andreas transform fault.

Exercise 1

Copy out the crossword and then attempt to solve it. (*Solution on p. 142*)

Across

1 Sediments are scraped off the surface of the subducting plate to collect in ________ (5)
4 Convection currents in the ________ cause the plates to move (6)
5 ________ is dissolved in a rising magma (3)
7 An island arc is a ________ of volcanic islands (3)
8 New rock is added to the ________ edge of the moving plates (4)

Down

1 A transform ________ happens where two plates slide past each other (5)
2 A dense oceanic plate ________ when it collides with a thicker plate (5)
3 Oceanic plates return to the mantle because of ________ convection currents (4)
4 Plutons of ________ rise towards the surface (5)
6 Buckling and folding happen when ________ plates collide (3)

Exercise 2

Write out and complete the following statements using the words listed below.

1 Plate tectonics is the theory that the Earth's surface is divided into a number of rigid ________ called plates.
2 The plates slowly ________ across the Earth's surface, in response to ________ in the plastic mantle below.
3 There are few earthquakes or volcanoes in the ________ regions of plates. Most deforming changes take place at plate ________.
4 The name given to a crack along which rocks move or slide is a ________.
5 Folding, faulting, volcanoes and earthquakes are all the result of ________ processes.
6 When two plates move apart the rocks are stretched by ________ forces.
7 When two plates ________, the rocks buckle and fold as they are compressed.
8 Plates are moving at about ________ on average.

collide tectonic fault move central 6 cm/year
convection edges tensional segments

Exercise 3

1 Study Fig 2.3.
   a) Name **two** plates composed entirely of oceanic material.
   b) Where are there more earthquakes – along ocean trenches or ocean ridges? Why?
2 What kind of plate motion **adds** new crustal surface area to the earth?
3 What is the name given to the faults produced by two plates sliding past each other?
4 What kind of motion results in the destruction of a plate?
5 What is the name given to molten rock below the surface?
6 What is the name given to the **zone** where new plate material is forming?
7 Explain why folded sediments are often found on continental land masses.
8 What are the longest and largest uplifted features on the Earth's surface?
9 Copy Fig 2.5.
   a) Why do faults develop in the first place?
   b) Why are fissure eruptions so common parallel to the rift?
10 Examine Figs 2.6 and 2.7. Explain fully how pillow lavas form.
11 Look at Fig 2.8. Explain how the island has formed.
12 What are magnetic stripes and how do they form? Why are they so important as evidence?
13 Explain how trenches, island arcs, earthquakes and plutons are created by subduction.
14 How are gulfs and inland seas formed?

3 THE CHANGING EARTH

Throughout its history the Earth's surface has undergone many changes. One of the biggest changes which is still going on is the movement of the continents. The theory of *continental drift* is not particularly new. For many years it was rejected as a wild idea. It is now much more widely accepted because of the evidence provided by plate movements and sea floor spreading.

CONTINENTAL DRIFT

About 200 million years ago all the continents were part of one vast land mass, called *Pangaea*. This supercontinent eventually began to break up and two land masses formed, a northern one called *Laurasia*, and a southern one called *Gondwanaland* (see Fig 3.1). This was the world of the dinosaurs.

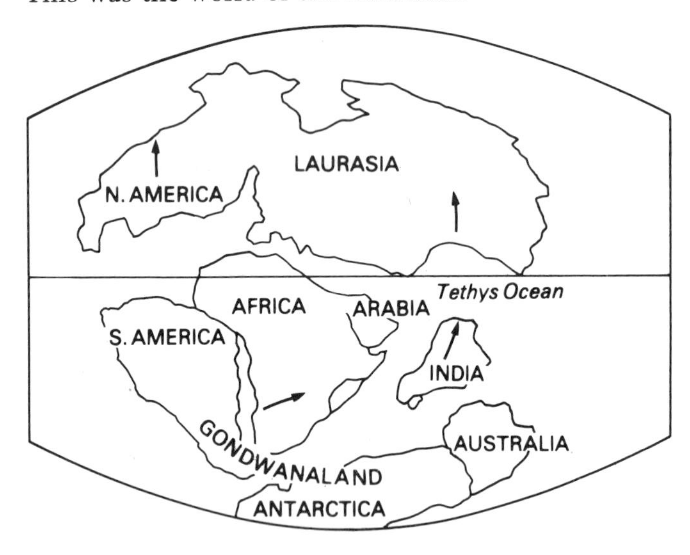

Fig 3.1 Early Cretaceous – 135 million years ago

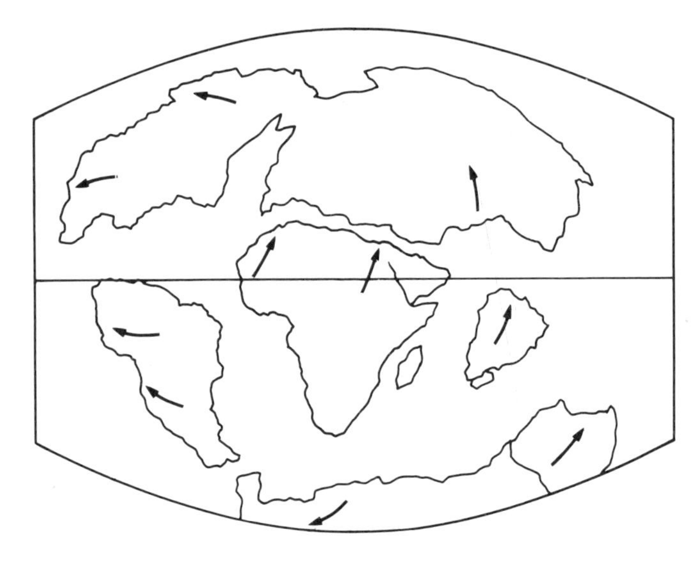

Fig 3.2 Early Tertiary era – 65 million years ago

Fig 3.1 shows how the northern continent began to split up as North America moved westward. The South Atlantic Ocean is just beginning to open as South America and Africa move apart. India has separated from the southern mass and has begun drifting northwards along with Africa and Arabia. The Tethys Ocean is closing.

The situation in the Tertiary Era (60 million years later) is shown in Fig 3.2.

The idea that the continents were once joined together mainly came from matching up the shape of facing coastlines. You can test this idea for yourself.

Activity 1: Continental Jig Saws

Trace the shapes of the southern continents onto card. Then cut them out and find the best way to fit them together into a single continent. When you have decided on the best fit, make up an outline map of Gondwanaland, placing each card on paper and drawing round it. Compare your finished map with the map in Fig 3.3.

The positions of the southern continents in Gondwanaland was not decided simply on the basis of the best fit of their coastlines. Other evidence was taken into account.

The Gondwanaland glaciation

Late in the Carboniferous period (300 million years ago), all the southern continents were in the grip of an intense ice age. There was a thick, slowly-moving ice sheet, which was responsible for transporting and depositing huge amounts of rock material, known as *till*. Till of this age is found in all the southern continents. It is even found in India north of the equator.

It is difficult to explain how ice could have spread so far north with the continents in their present fixed positions. It seems more likely that the southern

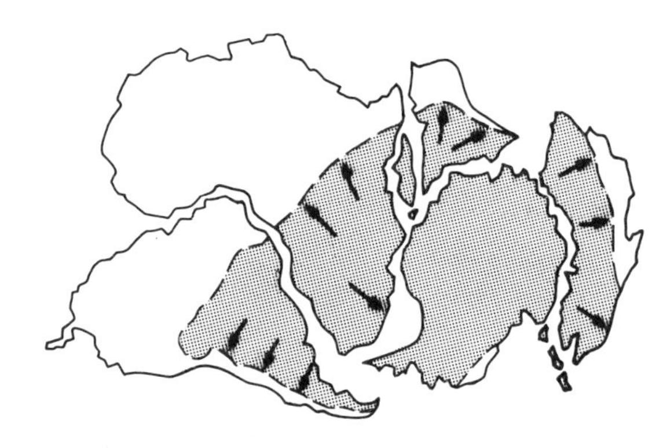

Fig 3.3 Reassembly of the southern continents – Gondwanaland. The shaded area was covered by a thick ice sheet, late in the Carboniferous period. (→ = direction of ice flow)

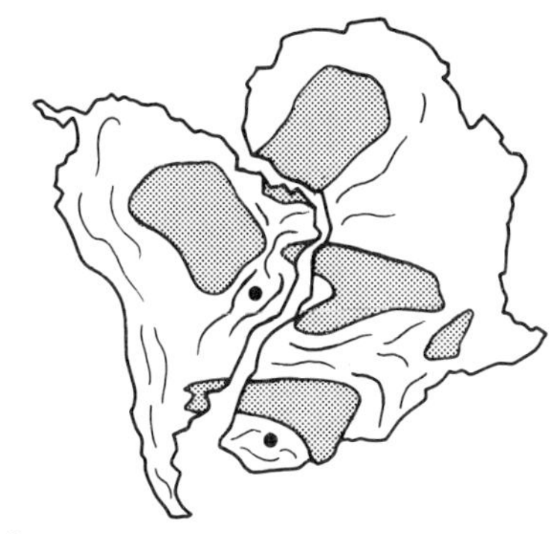

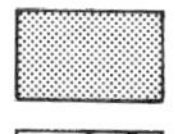

Fig 3.4 Matching structures (see text for explanation)

continents were once grouped much closer together near the South Pole. They have since drifted apart (Fig 3.3).

Matching structures

Many of the continents with matching facing coastlines also show matching structures. For example, rocks of the same age, type and structure in South America match up with the same rocks in Africa six thousand kilometres away (Fig 3.4).

Fossils

Fossils are the remains of animals or plants preserved in rocks. *Mesosaurus* was a small freshwater reptile about 50 cm long. Its fossil remains are found only in lake bed deposits of early Permian age 270 million years ago in South America and Africa. It would seem that the lake area it lived in must have split, and the two halves drifted thousands of kilometres apart.

Glossopteris was a Permian fern. Its fossil remains are found in every southern land mass but *not* in any of the northern continents. If the continents had always been where they are now, how was this plant able to spread right across the southern continents? This suggests a single southern land mass in the Permian period.

Coal seams

The evidence for warmer climatic conditions in the Permian is provided by fossils, like *Glossopteris*, and by the presence of thick coal seams. Layers of coal are known to form in swampy, wet, forested areas. Coal seams of the same age are found in all the southern continents.

Polar magnetic evidence

We saw in Chapter 2 how the magnetic record held in certain rocks can be useful evidence for sea floor spreading. It is also possible to identify the location of each continent. Using the magnetic record, the location is defined as the distance from the poles (*ie latitude*) at any given time in the past (see page 12). When the method is applied to the southern continents, it shows that they were all positioned very close together from the late Carboniferous period to early Cretaceous times – a total period of about 150 million years. Then they began to separate.

The value of this evidence is that it confirms the findings based on other evidence, and is of a completely different nature.

GEOGRAPHICAL CHANGES

Closing oceans

The slow movements of the continents can cause remarkable changes on the Earth's surface. For instance, colliding continents can cause uplift of fold mountains, and this in turn can cause major changes in climate. Plate movements ensure that today the Pacific Ocean is getting steadily smaller and *closing* whereas the Atlantic is opening. Consider also the case of the Tethys Ocean, at one time as wide as the Atlantic is today. The northward drift of India and Africa caused the Tethys to disappear, although remnants of it are seen as landlocked seas, eg, the Black Sea and the Mediterranean. These seas are floored by thinner oceanic crust.

About 10 million years ago, the northward movement of the African plate against Europe cut off the flow of water between the Mediterranean Sea and the Atlantic Ocean at Gibraltar. Over the next 1000 years the Mediterranean disappeared completely. The seawater steadily evaporated, leaving salt deposits across a massive Mediterranean lowland plain. Former islands such as Cyprus, became mountains standing high above the plain. Then about five million years ago the Atlantic Ocean again cut through the rock barrier. A huge waterfall of seawater several miles wide must have existed for several years near Gibraltar as the basin filled up again to reform the Mediterranean Sea.

Opening oceans

When two continents move in opposing directions, an *opening* ocean basin develops between them. The Atlantic Ocean is a good example of an ocean that is widening. The North Atlantic Ocean is about 6000 km wide, from Cape Cod to the Canary Islands. It is widening at a rate of three centimetres per year; so in one million years it widens by 30 km. In the 500 years since Colombus discovered America, the Atlantic has widened by 15 m. Using this information, can you work out the date at which the Atlantic first began to open?

A new ocean basin is produced by the formation of a long fissure in a continental mass. The crack slowly widens as tensional forces pull the two halves of the continents apart. A rift valley develops and lakes form

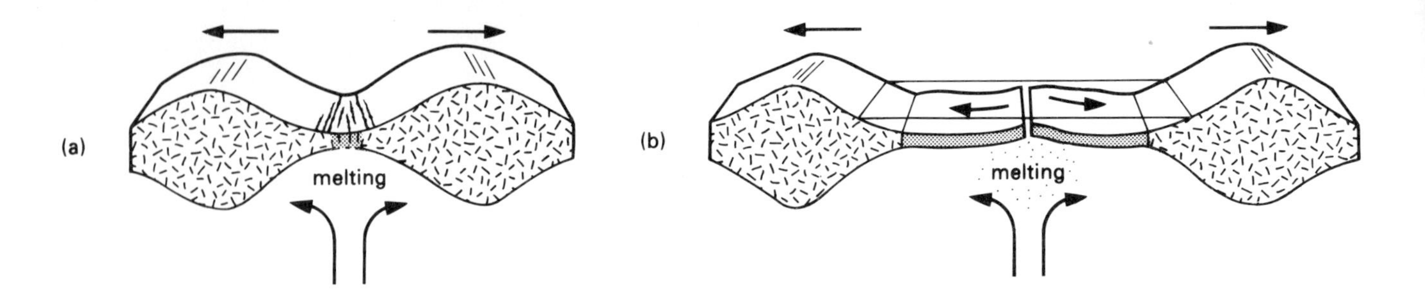

Fig 3.5(a) Continental crust thins and a rift valley develops. The rift widens as magma is injected from the mantle. (b) The rift valley is flooded by the sea. Sea-floor spreading adds new material to the back edge of each continent, as they move apart.

on the valley floor. This is happening in east Africa at the moment. As the rift valley widens, the lowland fills with sea water. This forms a gulf and eventually a narrow sea (Fig 3.5).

Twenty million years ago Arabia was joined to Africa. The Red Sea and the Gulf of Aden are opening ocean basins, floored by new oceanic crust (Fig 3.6). The rift valley now runs along the crest of a spreading ridge on the Red Sea floor. There is a high heat flow along the crest of the ridge. New plate material is forming. The mid-Atlantic ridge with its rift valley, marks the original rift between the Americas, and Europe and Africa. The continents have moved away from the rift as new oceanic plates have formed.

One of the most recent geographical changes is the formation of the narrow land isthmus between the north and south American continents (Fig 3.7). This narrow strip of land is the result of uplift of a chain of volcanic islands, similar to the West Indies island arc in the Caribbean. There is also folding of associated sediments. This complicated process is mainly caused by the subduction of the Cocos plate beneath the Caribbean plate. A relatively small geographical change like this could have untold effects. For instance, it could be responsible for changes in the circulation of ocean currents. The warm Atlantic current known as the Gulf Stream has been pushed north east towards northern Europe. Britain's present temperate and mild wet winter climate could therefore be linked indirectly with plate movements.

THE EFFECT ON LIFE

Animal life has developed differently on isolated continents. South America was cut off from the rest of America throughout Tertiary times. A study of fossils shows that 29 mammal groups unique to South America developed, among them the armadillos, sloths and ant-eaters. A group of 27 different mammals developed in North America. Then the Panama isthmus linked the two Americas three and a half million years ago and allowed the animals to migrate north and south. Most of the South American animals could not compete with the meat-eating invaders from the north and they died out. One exception was the armadillo which survived and successfully moved north. In a short time both continents were populated by the same 22 families of mammals.

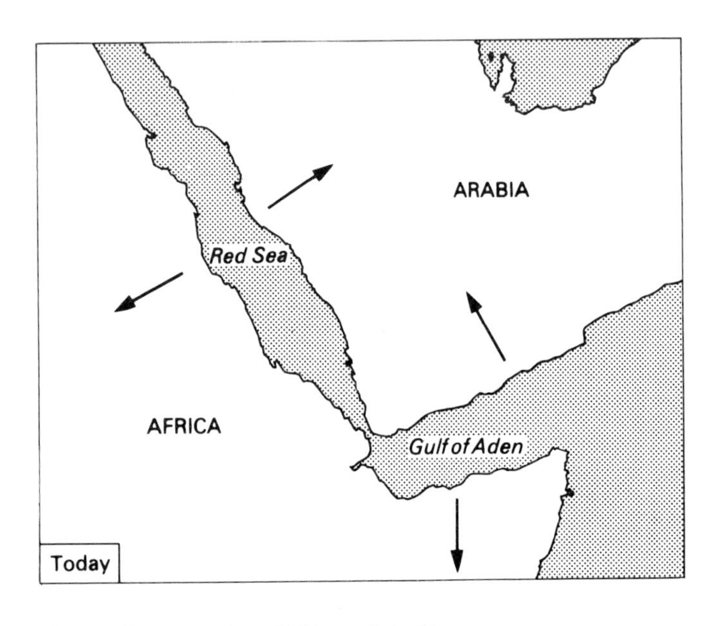

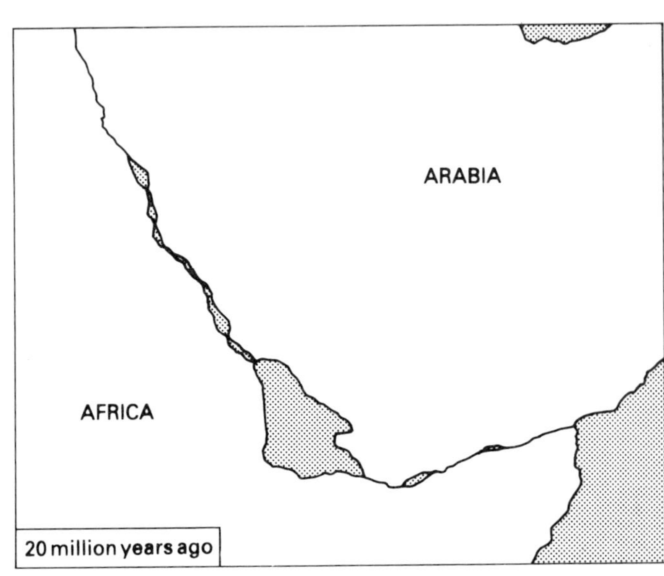

Fig 3.6 The separation of Africa and Arabia

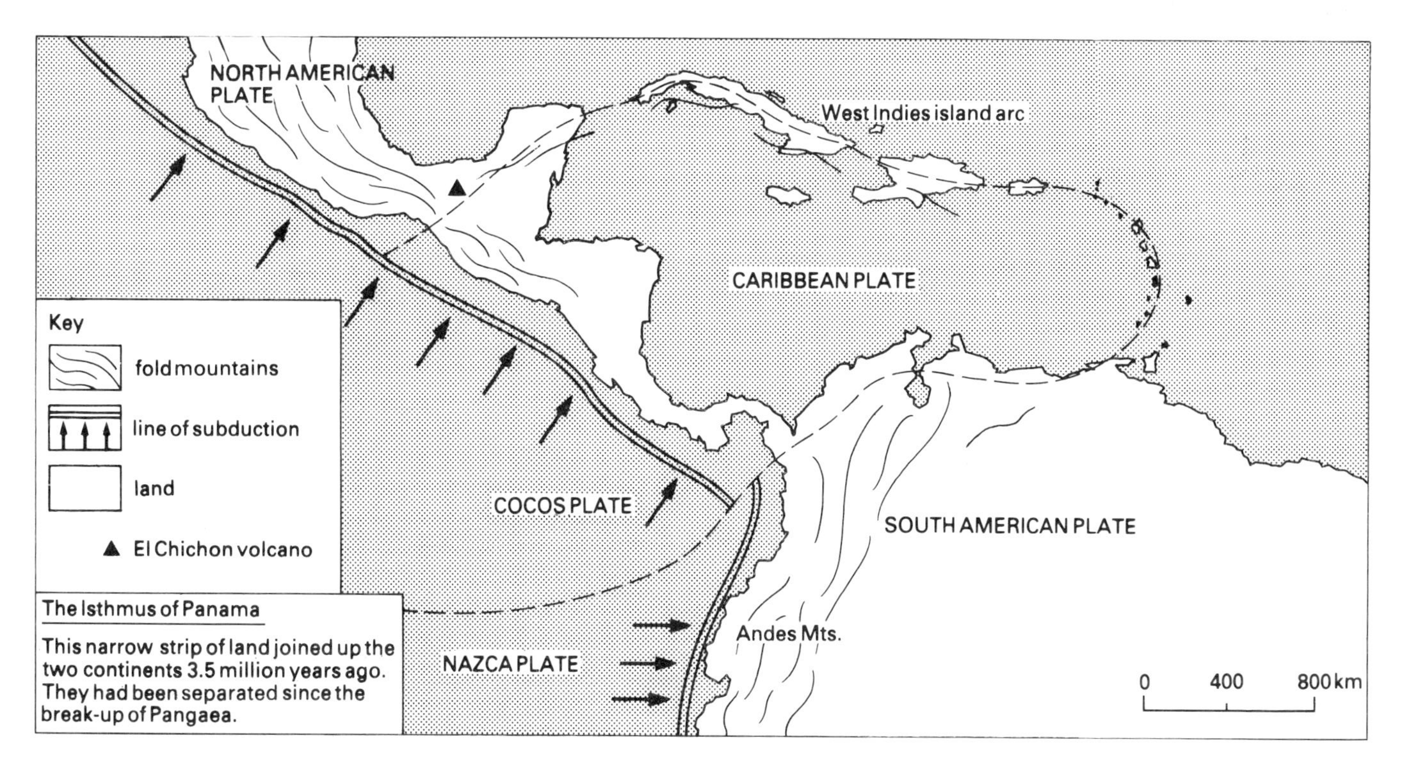

Fig 3.7 The Isthmus of Panama.

Exercise 1: Continental drift

Copy out the crossword square and then attempt to solve it. (*Solution on page 142*)

Across

4 Another name for a layer of coal (4)
6 *Glossopteris* was this kind of plant (4)
7 This deposit is important evidence of an ice age (4)
9 The name of a chain of islands, like the West Indies (3)
11 Volcanic activity produces a high flow of this at the Red Sea ocean ridge (4)
13 The opposite direction to Gondwanaland's position (5)
14 A period of time (3)
15 Many coal seams like this have been found in the southern continents (5)

Down

1 Continental ________ is a theory about movement of land masses (5)
2 The Black Sea is a ________ of a much larger ocean (7)
3 More successful animal predators from a previously isolated continent do this to their prey (4)
5 It is possible to ________ structures in Africa and South America (5)
8 The Gondwanaland glaciation was in ________ Carboniferous times (4)
10 Gonwanaland's position was ________ of the Equator (5)
12 Fig 3.2 shows where the continents were in the ________ Tertiary (5)

Exercise 2

Write out and complete the following statements using the words listed below.

1 All the continents were once grouped into a single supercontinent called ________.
2 One hundred and thirty five million years ago, it was a different world for the ________.
3 Fossils of Mesosaurus are found only in ________ deposits in Africa and South America.
4 ________ conditions are ideal for coal formation.
5 Magnetic evidence is of value in confirming earlier findings about the ________ of continents.
6 The Pacific is a ________ ocean, whereas the Atlantic is still ________.
7 The Mediterranean area was once a ________ plain.
8 Ships following the same route that ________ took, now have to travel 15 metres further.

9 Ocean basins begin when a long _______ appears in a continent.
10 The American continents are now linked by a narrow _______.

lake Pangaea closing dinosaurs positions Columbus swampy isthmus lowland rift opening

---

Exercise 3

1 Describe the main changes in the position of the continents between the Cretaceous and Early Tertiary (see Figs 3.1 and 3.2).
2 Summarise the main lines of evidence that confirm the theory of continental drift.
3 Describe the processes causing the two continents to drift apart in Fig 3.5.
4 Copy Fig 3.6 and use it to explain why the Red Sea is called an opening ocean.
5 a) What is meant by a closing ocean? Give at least *two* examples.
   b) Describe the processes involved.
6 What effects have plate movements had on the climate and conditions in the Mediterranean area?
7 Look at Fig 3.7, the Isthmus of Panama.
   a) Explain the processes that have formed this land-bridge.
   b) Describe the effects of continental movements on land and animal life in this area.

---

4 MOUNTAIN BUILDING

The highest land mountain in the world is Mt Everest. The rocks near its summit are sediments from the sea bed. They contain fossil sea shells, now lifted 8840 m above sea level. Movements of plates during the last 30 million years have produced the remarkable uplift of these rock layers and the fossils they contain. Mt Everest is in the Himalayan range. The range is made mostly of parallel lines or 'chains' of folded and crumpled layers of rock, that were once flat on the sea floor. Similar fold mountain chains can be seen in other parts of the world and are shown in Fig 4.1.

WHAT ARE FOLDS?

Rock layers become folded because of sideways compression (Fig 4.3). In many cases this happens as sea-floor sediments become trapped between two colliding plates. This effect can easily be demonstrated (see Activities 1 and 2).

Activity 1

Place two heavy books some distance apart on a cloth-covered table. Slide the two books together. Notice what happens to the tablecloth.

Activity 2: Plasticine folds

Squeeze horizontal layers of different coloured Plasticine into folds as shown in Fig 4.2. Sketch the results.

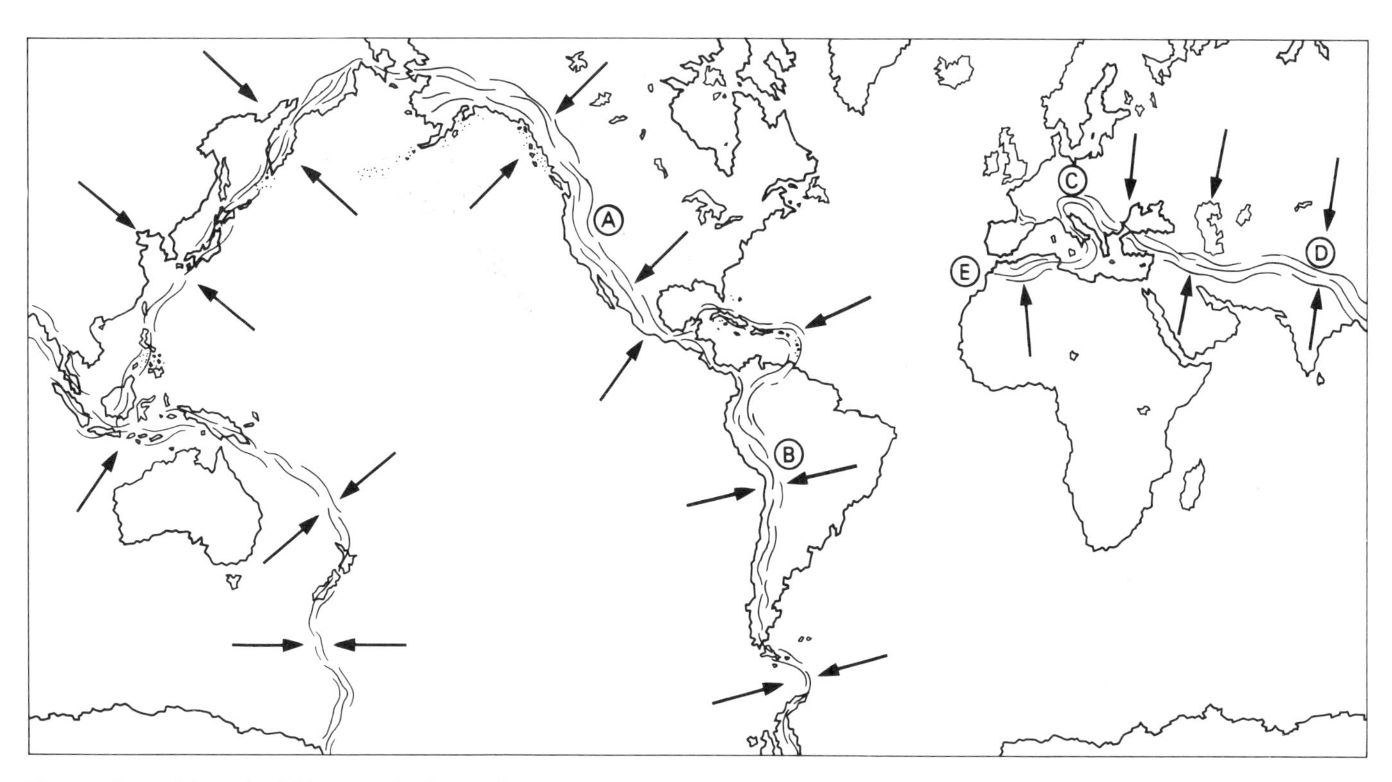

Fig 4.1 The world's major fold mountain chains all occur at active plate edges, and are often associated with earthquakes and volcanoes. The arrows show where plates are in collision. (Question 1, Exercise 3 refers to letters A–E.)

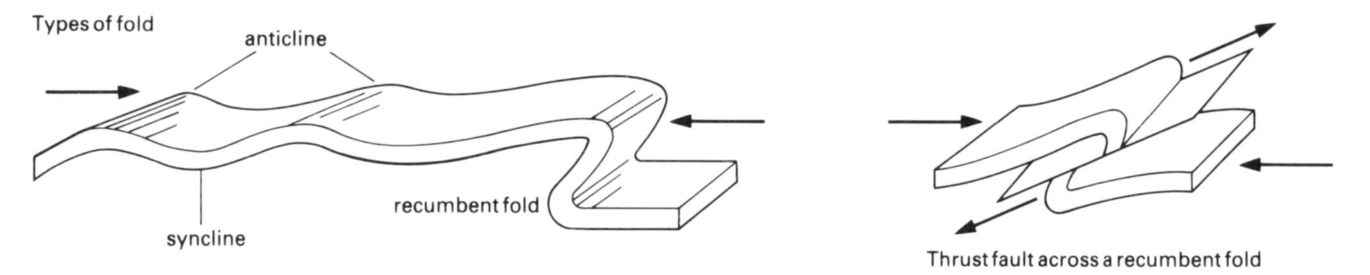

Fig 4.2 Types of fold mountains

Fig 4.3 Folded rock layers in the Rocky Mountains, USA

The main kinds of fold are shown in Fig 4.2. An *anticline* is an upfold. Downfolds are known as *synclines*. In most mountain belts *recumbent* overfolds are also found. They are often *faulted* – the sideways stress has been so great that the fold is broken along a *thrust fault*. The top part of the fold is then pushed across the bottom part.

MELTING AND ALTERATION OF ROCKS

Mountain ranges occur in areas of extreme crustal disturbance at the edges of plates. Many active volcanoes are found in fold mountain areas, together with injected plutons of liquid magma. The rising plutons help to cause uplift as they dome up the sediment layers (see page 23).

In addition the folded rocks of a mountain range undergo much heat and pressure which can alter their structure. Rocks that were sediments become 'baked', and re-form with new crystals. This alteration is known as *metamorphism*.

Uplift of the Himalayas

The formation of the Himalayas is due to the northward movement of the Indian plate in the last 100 million years, with the subduction of the oceanic plate in between (Fig 4.4). Sediments lying on the Tethys Ocean floor between the Indian and Asian land masses have been compressed into folds, and the ocean itself has disappeared. The oceanic plate has also been destroyed by subduction, and the less dense sediments scraped off and folded. India is still grinding its way northward into the Asian continent at a rate of four centimetres per year and the mountains are still being uplifted.

At the moment there is in effect a double thickness of over 70 km of continental crust in the Himalayan mountain range. Eventually the two continental plates will lock together because it is impossible for light continental crust to be subducted into the mantle. The continents are behaving like two giant corks; their combined buoyancy is able to support the weight of the overlying Himalayan folds. As the two continents lock together the Indian land mass will eventually cease to push northward. Then the oceanic crust in the south will break (Fig 4.4) and start to subduct below India. In time, a new range of mountains could form, and somewhere near India's southern coast, Indian Ocean sediments will be scraped off and folded.

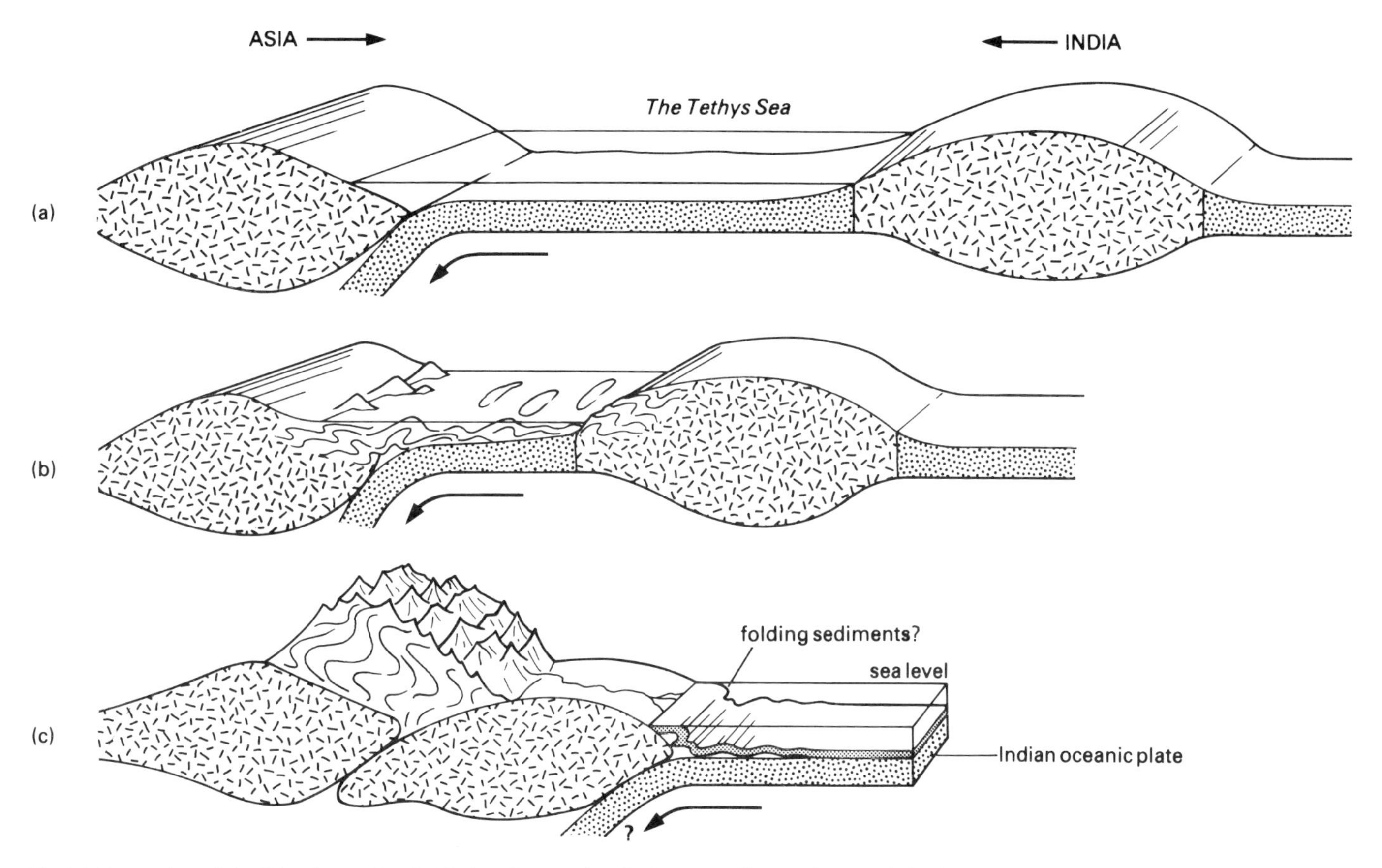

Fig 4.4 Formation of the Himalayas. As the Tethys Ocean closed, the ocean-floor sediments folded. Eventual collision of the two continental landmasses caused the uplift of the Himalayas. Sediment provided by erosion of the Himalayas spread out on the Indian plate. The oceanic plate may start to subduct below the Indian continent as the sediments are added to the southern edge.

THE ANDES

The Andes is a range of fold mountains forming at the leading edge of the South American continent. As the South American plate spreads westwards, it collides with the Nazca plate. In this situation a thicker continental plate is meeting a thinner and denser oceanic plate. The thinner plate sinks at an angle below the South American continent. The low density sea-floor sediments are scraped off and added to the continent (Fig 4.5). At the same time there is some

Activity 3: Making fold mountains

Press a thin layer of modelling clay (eg, Play-doh) onto the surface of a flat-bladed flexible baking knife. Then slide the blade underneath a firmly-held chopping board (Fig 4.6). Note how the layer of clay folds and separates from the descending knife blade. In this activity, the chopping board is the continental plate; the baking knife is the subducting oceanic plate and the layer of modelling clay is the low density sediment.

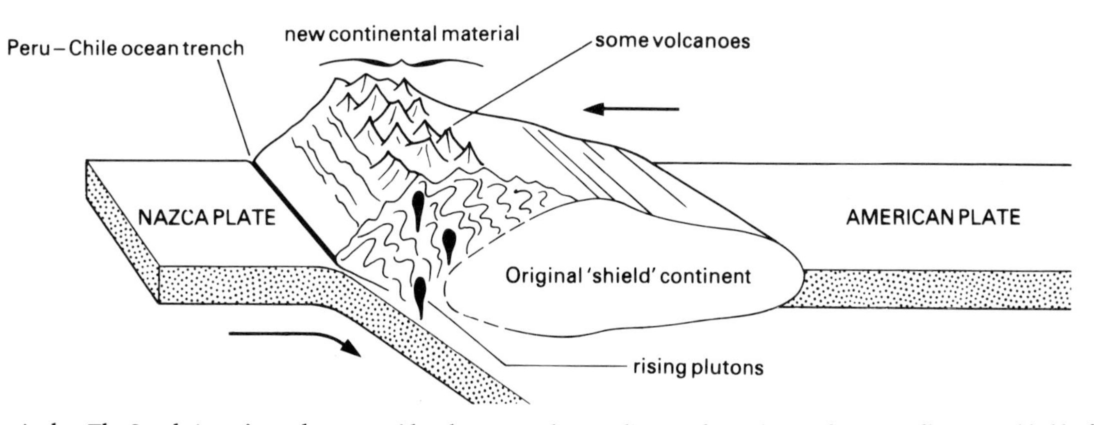

Fig 4.5 The Andes. The South American plate over-rides the Nazca plate. Sediments from the Pacific Ocean floor are added by folding onto the continental edge, to form the Andes range. Compare this with the situation shown in Fig 2.10. What are the main differences?

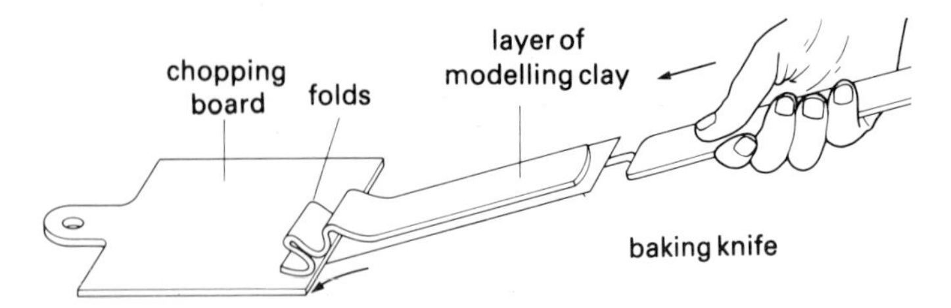

Fig 4.6

melting of the top of the oceanic plate as it sinks into the mantle. This is caused partly by friction, and partly by the presence of water which lowers the melting point. Magma is produced which rises as plutons. This helps to cause uplift of the sediment and, if the magma reaches the surface, volcanoes are also formed.

It is possible to illustrate how layers of sediments can be buckled and folded by collision, using everyday materials (see Activity 3).

THE FUTURE

Since the plates are still moving, and the continents still drifting, it follows that mountains will still be uplifted millions of years in the future. North and South America will again separate. East Africa will be split along the East African Rift Valley and begin to drift eastwards. The Black Sea and Caspian Sea will disappear. New mountains will be formed in this area, as the Arabian plate pushes northwards into the Eurasian plate.

Eventually, perhaps, a new supercontinent could again form as the Americas move towards China and Australia, and the Pacific Ocean closes. Today's huge fold mountains will by then have been eroded away. The layers of sediment they produce will form the fold mountains of the future.

Exercise 1: The mystery mountains

(Solution on page 142)

Solve the five clues and discover the name of the mystery mountains.

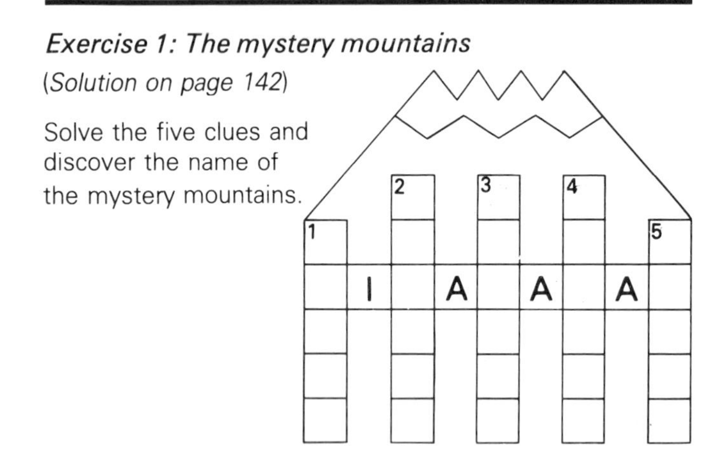

1 Another name for a mountain range.
2 The highest point on a mountain.
3 How folds of sediments form mountains.
4 The term used to describe a certain thickness of sediments.
5 India collided with this continent.

Exercise 2

Write out and complete the following statements using the words listed below.

1 Mt ________ is 8840 m above sea level.
2 The Himalayas have formed in the last ________ years because of plate movements.
3 The fold mountains of the world are all positioned at ________ plate edges.
4 An anticline is an ________ and a syncline is a ________.
5 An ________ is called a recumbent fold.
6 Sea floor sediment becomes squeezed between two ________ plates.
7 Metamorphism is the alteration of ________ by heat and ________.
8 India is still pushing into the ________ continent at a rate of 4 cm/year.
9 It is impossible for ________ crust to sink into the mantle.
10 A pluton is a rising bubble of hot ________.

upfold active Everest overfold rocks colliding downfold magma pressure continental Asian 30 million

Exercise 3

1 a) Make your own copy of Fig 4.1, using a world outline map.
b) Use an atlas to find out the names of the fold mountain ranges labelled A–E.
2 Describe the three main kinds of fold and explain how they form.
3 What is meant by a 'thrust-fault'?
4 Describe some of the other processes, apart from the folding of sediments, that are found in mountain ranges.
5 Study Fig 4.4 carefully and then explain fully the processes involved at each stage in the formation of the Himalayas.
6 What are the main differences in formation between
a) the Himalayas and
b) the Andes?
7 'Continental material can never return to the mantle'. Discuss the evidence for this statement.

5 EARTHQUAKES

Most of the world's earthquakes happen at the edge of the plates as they constantly grind against each other. The plates often meet at a fault zone. As they move, major shock waves are produced (Figs 2.2 and 2.3, page 10). The main direction of movement of the plates in the Turkish area can be seen in Fig 5.1.

In the past 45 years there have been over 20 major earthquakes, resulting in the death of over 60 000 people, as the Turkish plate shudders and jerks its way westward along the Anatolian fault. The rate of movement is about eight centimetres per year. The large African plate and the Arabian plates are pushing northwards against the southward movement of the large Eurasian plate. The Turkish and neighbouring Aegean plates are caught in the middle of this large-scale crushing movement.

The violent shaking of the ground which occurred in the Erzurum event, and destroyed so many buildings, is a common feature of all earthquakes. The vibration can be so severe that the ground is thrown into rolling waves. Long cracks open on the wave crests and close soon afterwards as the ground falls. Observers tell of parked cars bobbing up and down, and potatoes flying out of the ground, propelled by the force of the shock. The effect of such motion on buildings is catastrophic (Fig 5.2). Constructions such as bridges, railways, gas and water pipes are very prone to destruction. Falling masonry is usually the chief cause of human deaths. However, in the 1906 San Francisco earthquake the

Children Survive Earthquake but Freeze to Death

3 November 1983 – Erzurum, Turkey

The death toll from the Erzurum earthquake may reach 2000. Survivors face their third night in the open. The severe winter has begun and there are reports of many of the homeless children freezing to death.

In the village of Koprukoy an old man said: "We don't need food, we need shelter and shovels to bury the dead. My house has gone, my daughter and my grandchildren are dead. So many of the bodies are our children."

Some food, tents and blankets are beginning to reach the devastated area but communications are difficult. Roads have been destroyed by landslides and there is a severe shortage of petrol for vehicles. More than 50 towns and villages are almost totally destroyed. Relief organisers who know the bleak plateau in winter are afraid that relief efforts will be stopped soon, because of the heavy winter snows. Homeless refugees are converging on Erzurum, the provincial capital. The centre of the earthquake, which registered 7.6 on the Richter scale, was 35 miles to the east of Erzurum. The buildings of the city were damaged but there were no casualties.

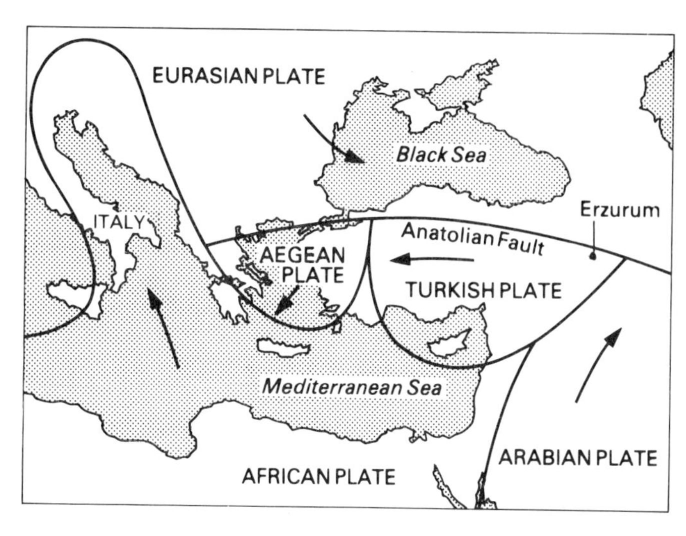

Fig 5.1 Plates and their movement, in the eastern Mediterranean area

Fig 5.2 Damage in the suburbs of Anchorage, Alaska, USA, resulting from the earthquake in 1964

greatest loss of life was due to fires resulting from ruptured gas mains. The fires raged unchecked for days. They were difficult to put out because all the water mains were ruptured.

HOW COMMON ARE EARTHQUAKES?

There may be up to 200 000 earthquakes every year. Most of these are minor tremors that can only be detected using sensitive instruments. Only around 20 earthquakes a year are really severe. They are mostly in remote unpopulated regions. They pass largely unnoticed and often are not reported by the media.

MEASUREMENT OF EARTHQUAKES

The extent of an earthquake can be measured by looking at how much damage there is. The effects are compared to an *intensity* scale, first devised by the Italian scientist G. Mercalli in 1902. It is also possible to draw lines of 'equal destruction' around the centre of an observed earthquake. These are called *isoseismal lines* (Fig 5.3).

In 1935 C.F. Richter devised a scale to show the size or *magnitude* of an earthquake. This scale is based on the total energy released at the source, as recorded on instruments called *seismometers*. Both the Mercalli and Richter scales are shown in Table 1. Comparisons between the two scales are approximate.

WHAT CAUSES AN EARTHQUAKE?

Any sudden release of energy can cause the ground to vibrate. This can be caused in an artificial way, for example, by dynamiting a quarry rock face, or by underground testing of nuclear bombs. The energy released by a major earthquake (8.6 Richter) is roughly equal to the energy released in a 100-megaton nuclear explosion – that is 5000 times the power of the bomb that flattened Hiroshima.

The fact that earthquakes happen so often is evidence that the crustal rocks are under constant stress. If the rocks are distorted beyond the limit of their strength, then they will suddenly fracture and release the stored energy. This is true of any material. For instance, a ruler is stressed when it is slowly bent. As it is bent energy is stored. When the ruler snaps, all the energy is suddenly released.

A sudden new fracture, or a movement along an existing fault, produces *seismic waves* (shock waves). The *focus* is the underground source of the earthquake. The *epicentre* is the nearest point at the surface to the source of the shock. Fig 5.3 shows seismic waves radiating from the focus and their relationship to isoseismal lines. The most intense vibration is at the epicentre with lessening disturbance further away. The Roman numerals refer to the Mercalli scale.

Big earthquakes along transform faults have focal

Table 5.1 Corresponding magnitudes (Richter)

MERCALLI Intensity (degree of shaking)	Description of characteristic effects	RICHTER Magnitude (total energy released)
I	Instrumental: detected only by seismographs	2
II	Feeble: felt only by sensitive people	3
III	Slight: like the vibrations due to a passing light lorry	
IV	Moderate: like the passing of a heavy road vehicle; rocking of loose objects, including standing cars	4
V	Rather strong: felt by most people; church bells ring	5
VI	Most people frightened; windows broken; dishes fall out of cupboards	
VII	Very strong: general alarm; walls crack; plaster falls	
VIII	Destructive: car drivers find it difficult to steer; masonry cracked; chimneys fall	6
IX	Ruinous: general panic; ground cracks appear and pipes break open	
X	Disastrous: ground cracks badly; many buildings destroyed; landslides on steep slopes	7
XI	Very disastrous: most buildings and bridges destroyed; all services (railways, pipes and cables) out of action; great landslides and floods; dams badly damaged	
XII	Catastrophic: total destruction; objects thrown into air; ground rises and falls in waves; cracks open and close	8

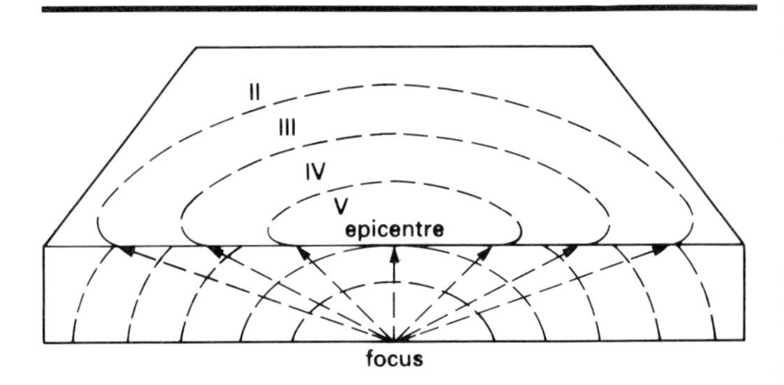

Fig 5.3 Isoseismal lines join together places of equal destruction

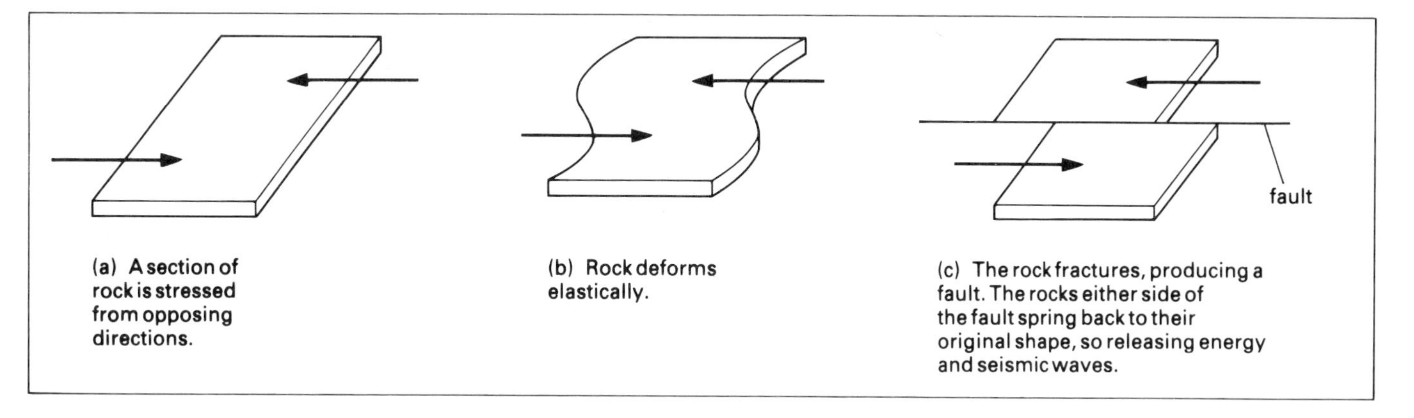

Fig 5.4 The pattern of movement on a transform (sideways-slip) fault, producing a shallow-focus earthquake

depths of 15–20 km. These are shallow-focus earthquakes, eg, along the Anatolian fault. Their pattern of movement is shown in Fig 5.4.

Very deep-focus earthquakes, up to 700 km, have been recorded. They are again caused by plate movements. In this case two plates are moving towards each other. The weaker plate slips downwards so producing earthquakes (Fig 5.5).

Recent earthquakes recorded on the Richter scale include the following:

Chile 1960	8.3–8.9
Alaska 1964	8.6
Muradiye, Turkey 1976	7.6
San Fernando, California 1971	6.6
Popayan, Columbia, 1983	5.5
Porthmadog, North Wales 1984	5.5

It is interesting to note that the Porthmadog earthquake was a deep-focus one, occurring along an old faultline. There was little surface damage and a Mercalli intensity of IV. The Columbian earthquake, on the other hand, was a shallow-focus one causing extensive damage and loss of life. The intensity measured up to VI on the Mercalli scale. The earthquakes were of equal magnitude.

The Mercalli scale is useful for making judgments based on observations of the intensity of the damage. However, the most accurate method of measuring earthquakes relies on the use of *seismometers* which measure the Richter magnitude, ie, the total energy released.

SEISMIC WAVES

Recording seismic waves: seismometers and seismographs

Seismometers are sensitive instruments designed to record seismic waves. They are capable of detecting faint earthquakes that would pass unnoticed to the average observer. Earthquakes thousands of miles away from the instrument can also be recorded.

There is a worldwide network of hundreds of seismological stations. Everyday dozens of earthquakes are recorded.

Types of seismometer

An early design was the rod seismometer. This was made of a series of cylindrical columns of varying height and thickness (Fig 5.6). A slight earthquake would dislodge the unstable tall columns leaving the shorter columns upright. This instrument could easily be used to estimate the *intensity* of an earthquake, but could not give the exact time nor duration of the earthquake.

A later design of seismometer is shown in Fig 5.7. You could make one like this yourself. A continuous timed recording is made on the rotating drum so the exact time and duration of the earthquake will be recorded.

Many of today's seismometers consist of a heavy magnet suspended by a spring. A coil is positioned

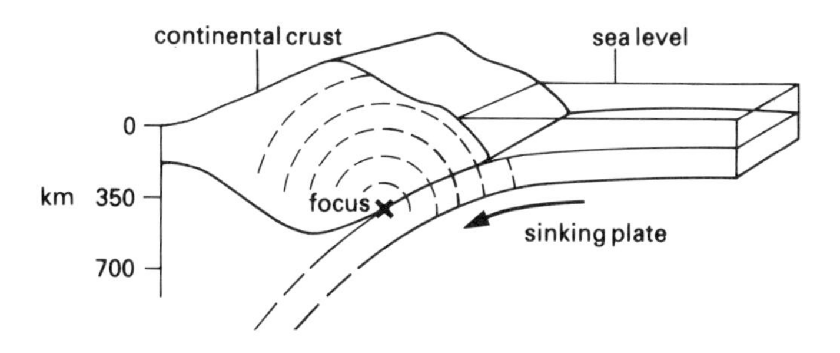

Fig 5.5 A deep-focus earthquake at a subduction zone

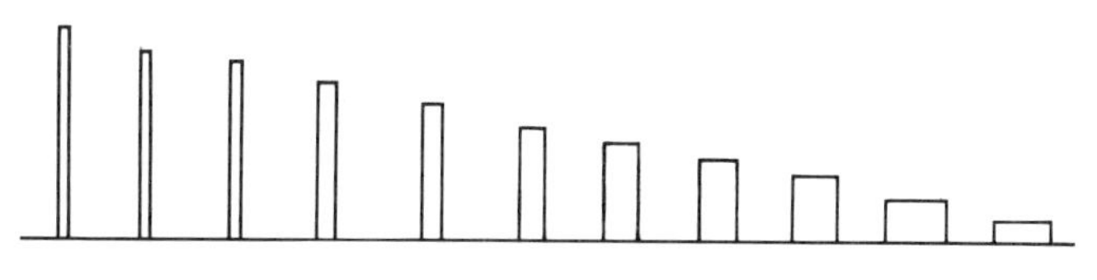

Fig 5.6 A rod seismometer

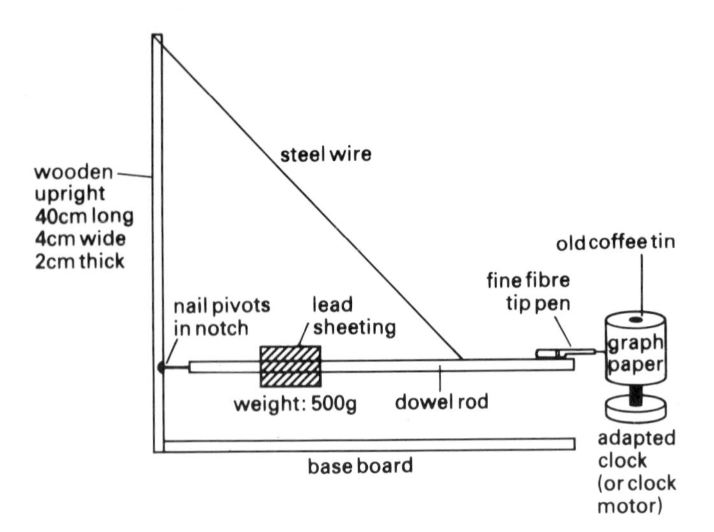

Fig 5.7 A model seismometer

Activity 1: Making a seismometer

It is a simple matter to construct a rod seismometer by using pieces of wooden dowel rod of varying lengths and diameter, following Fig 5.6. When you have prepared the rods, stand them up, eg on a desk top. To avoid any domino effect make sure they are far enough apart. Then simulate earthquake shock waves by banging the desk top with your fist.

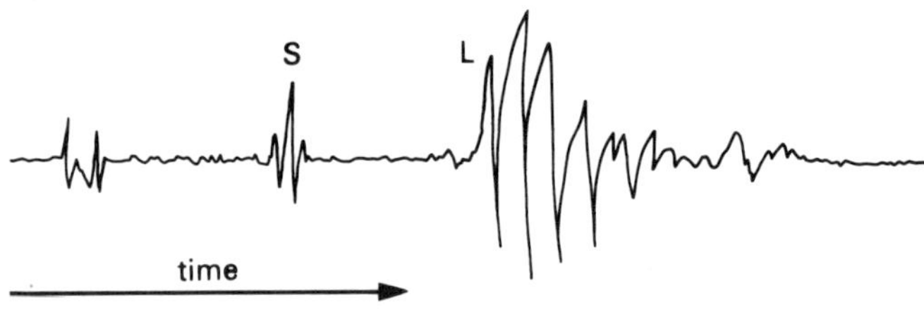

Fig 5.8 A typical seismograph reading

between the poles of the magnet. When the magnet vibrates up and down, an electrical current is created in the coil. This is then recorded as a *seismograph* (Fig 5.8).

How seismic waves travel

On the seismograph (Fig 5.8), you can see three distinct 'pulses' of shock, and their time-order of arrival. P, S and L waves travel at different velocities (speeds) because they are transmitted differently through the Earth.

P waves: are compressional waves. The rock material is first compressed and then stretched. Therefore, the motion of the rock particles is backwards and forwards. This movement takes place in the same direction as the direction of wave travel. The transfer of shock from particle to particle can be likened to a line of railway trucks being hit by a runaway engine.

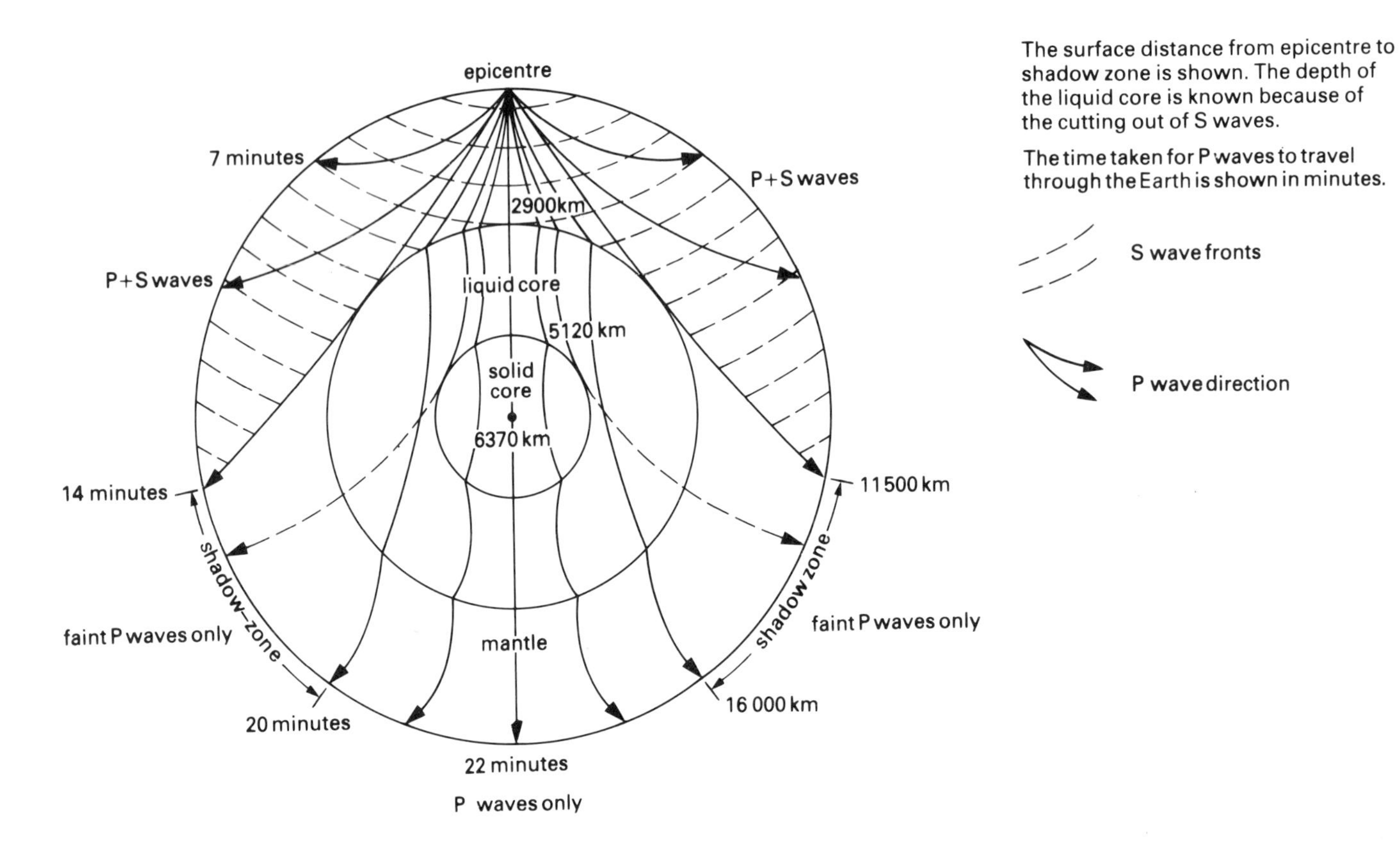

Fig 5.9 How P and S waves travel through the Earth

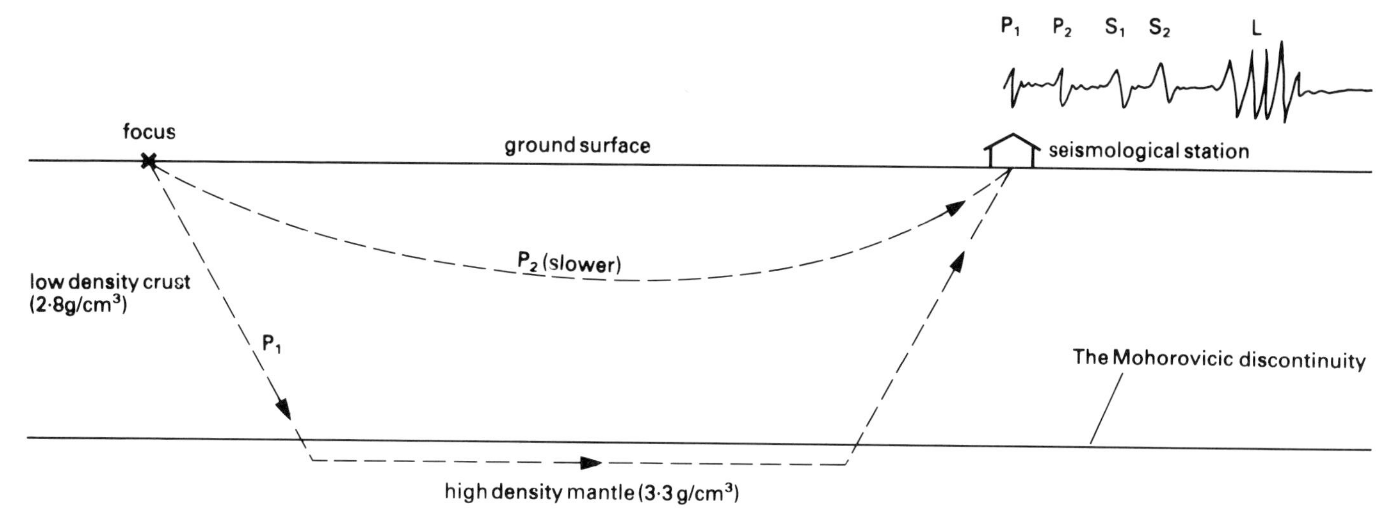

Fig 5.10 Discovery of the 'Moho' – the crust/mantle boundary

The force of the original impact is transferred through the line of trucks, as each truck collides with the one in front of it.

S waves: are transverse or shear waves. Their crests and troughs rise and fall at right angles to the direction of travel of the wave front. They can only travel through rigid materials, liquids cannot transmit them. S wave motion can be compared to the snake-like loop travelling along a taut skipping rope if one end is suddenly shaken.

L waves: are the slowest of the three types of seismic wave, and are confined to the Earth's surface layers. They have a rolling motion and cause the greatest destruction in an earthquake.

Using seismic waves to 'see' the Earth's core

P and S waves can be used to 'x-ray' the Earth's deep interior. When a major earthquake occurs, P and S waves pass into the Earth in all directions. Their velocities increase with depth because they are travelling through much denser rock. Both wave types are recorded returning to the surface, up to 11 500 km from the epicentre. After this point, S waves disappear entirely and P waves become very faint. Beyond 16 000 km from the epicentre, only P waves are recorded (Fig 5.9).

The zone where *body waves* (P and S) are absent or rare is known as the *shadow zone.* It is caused by the dense core acting like a lens. It bends the waves inwards. The disappearance of S waves, at 2900 km depth shows that the outer core is liquid. Very faint P waves appearing inside the shadow zone indicate a smaller inner solid core, which bounces the P waves outwards.

Using seismic waves to find the crust/mantle boundary

The boundary between the crust and the mantle was first discovered on 8 October 1909, when there was an earthquake near Zagreb, Yugoslavia. Dr A Mohorovicic (pronounced Mo-horro-vich-ich) noticed that the seismograph was rather strange; *two* sets of P waves and *two* sets of S waves had been recorded (Fig 5.10).

He suggested that the P_1 wave arrived quickly because of its passage through denser mantle rocks. The slower P_2 wave followed the more direct route through low density crust.

THE GEOLOGICAL EFFECTS OF EARTHQUAKES

Avalanches

Many of the world's steep mountains are in earthquake-prone areas. Earthquakes in mountainous regions can trigger off avalanches and landslides on the steep slopes.

The 1970 Peruvian earthquake set off one such avalanche. The people of the town of Yungay saw the avalanche start in the mountains high above them. Many started running towards higher ground. Within two minutes of the original earthquake, the avalanche tore through the town. The mixture of mud, ice and rocks buried most of the town. Only 92 of the town's 20 000 inhabitants managed to escape alive.

Tsunamis

Earthquakes may also create *tsunamis.* A tsunami is a giant sea wave. Once generated, the wave travels away from the epicentre, at high speed.

Tsunamis were produced in the 1964 Alaskan earthquake. There was a sudden displacement of water in the Gulf of Alaska. Giant sea-waves over 10 m high flooded the immediate coast of Kodiak Island carrying boats several miles inland. The wave fronts moved southwards down the coast of Canada and the United States. They still had heights of 7 m when they reached California. At Hawaii the waves were 2 m high, and 1 m waves were recorded in Antarctica. Fig 5.11 shows

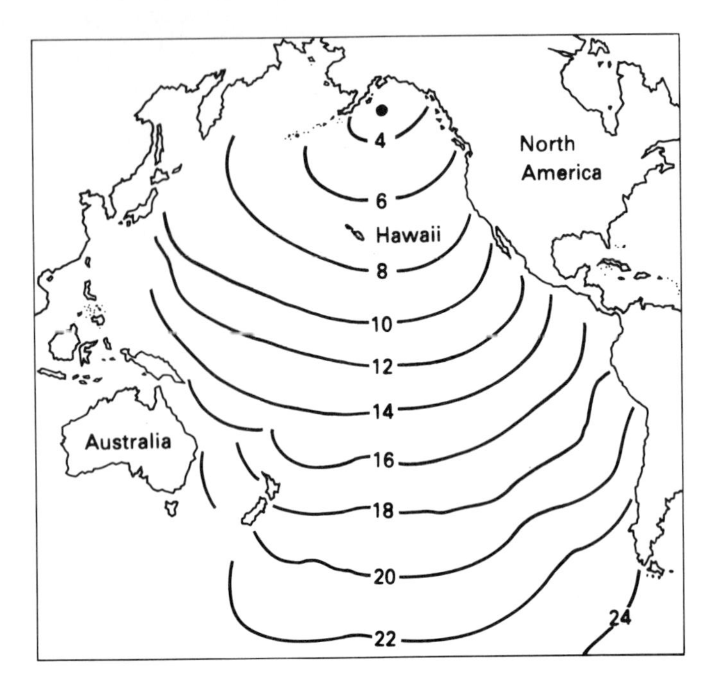

Fig 5.11 The time taken (in hours) for the Alaskan tsunami wave front to cross the Pacific Ocean, a distance of 15 000 km

the time taken for the tsunami to pass down the Pacific Ocean.

A tsunami early warning station for all countries surrounding the Pacific Ocean has been set up on Hawaii. Any report of tsunamis approaching is relayed to the countries likely to be affected. This has saved countless lives.

Tsunamis have wave lengths of hundreds of miles but their height in open ocean is no more than 1 m. The whole depth of water is involved in the wave motion. It is only when the wave reaches shallowing water than the wave height increases dramatically (Fig 5.12). In narrowing estuaries, tsunami can build up to heights of 20–30 m.

LEARNING TO LIVE WITH EARTHQUAKES

Many countries lie along plate boundaries and frequently have earthquakes. In Japanese cities there are earthquake drills where emergency procedures are practised. Other precautions, such as specially designed shock-proof buildings, can help save many lives.

The San Andreas Fault in California runs for 950 km from the Salton Basin in the south through Los Angeles and San Francisco. The fault is the boundary between two plates where the average movement is five centi-

Activity 2: Make your own tsunami

Cut out a wedge-shaped block of wood and place it at one end of a ripple tank to represent a sloping shoreline. Make waves and note the way in which wave height increases on the 'shoreline'.

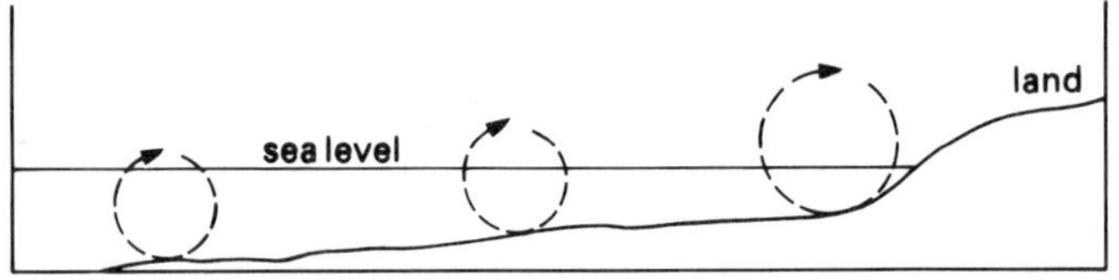

Fig 5.12 Changes in the wave height of a tsunami

metres per year. The rocks on either side spend many years in locked positions. As opposing stresses build up, there is a slow bending of the rock masses. Eventually the frictional strength is overcome and the rocks suddenly move to produce a major earthquake. The San Andreas Fault zone has *seismically active* sections where small earthquakes are frequent, and *locked sections* where there is no known recent movement. The strain builds up over many years in the locked sections, until there is a sudden 'break' (Fig 5.13).

The last big earthquake on a locked section was at San Francisco in 1906. It registered 8.6 on the Richter scale. Many parts of the city were completely flattened and about 700 people were killed. This part of the fault has not moved since. If the fault remains locked for 90 years (until 1996) then four and a half metres of sudden movement are possible (5 cm × 90 years = 450 cm =

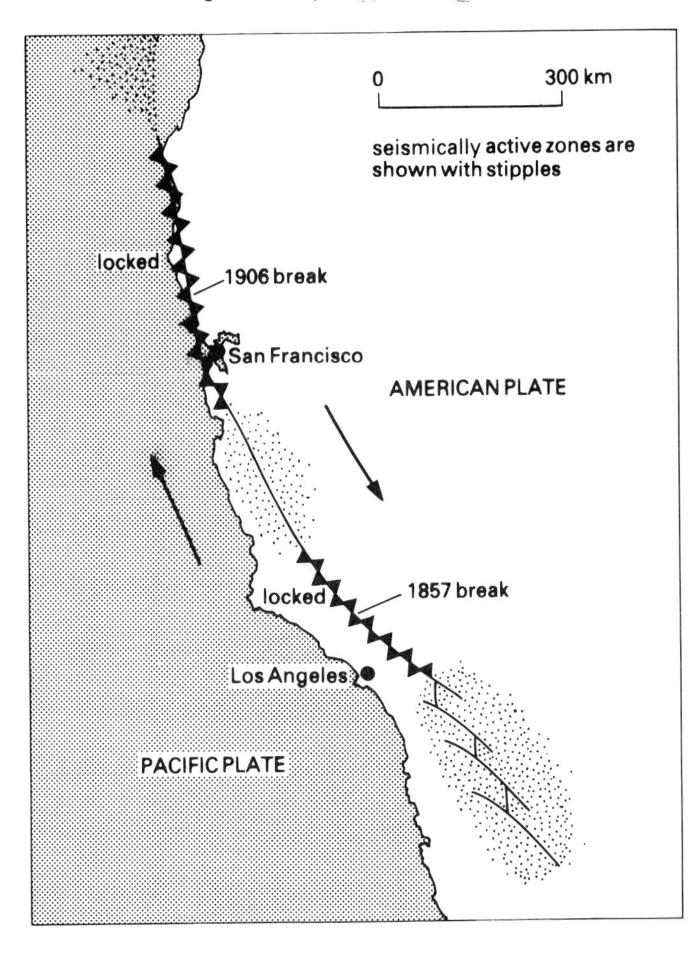

Fig 5.13 The San Andreas Fault

4.5 m). In fact a big earthquake can be produced by movement of less than 1 m. Since 1906, San Francisco has grown, with housing estates built astride the fault. There is now a total ban on building on the fault but many houses, schools and hospitals built before this ban are still there. There is little doubt that San Francisco will have another big earthquake soon.

Sensitive instruments to measure strain in the faulted rocks offer some hope of predicting a major earthquake and giving warning. The flexing and straining movements in locked sections of the San Andreas Fault can be detected using a *geodolite*. A laser beam is transmitted across the fault to a mirror. Here it is reflected back and recorded. The slightest strain of even one millimetre can be recorded.

However, methods of predicting earthquakes are not perfect. In 1976, the Chinese predicted a major earthquake in Peking. For several weeks seven million Chinese camped in the streets, much of the time in heavy rain. All office blocks were declared unsafe and businesses ground to a halt, but the earthquake never arrived. In the end officials declared the danger to be over and people returned to their normal lives!

CONTROL

One day it might be possible to control earthquakes along faultlines like the San Andreas. One idea is to deliberately set off a number of small earthquakes. This way the rock stresses would not build up to dangerously high levels, and the major earthquakes would be avoided.

Recent experiments have been carried out in the United States in fault zones similar to the San Andreas one. The method is to drill three boreholes, each three kilometres in depth, along the fault. Friction is increased in the outer borehole by pumping out as much water as possible. When this has been done, the water is pumped into the middle borehole. This reduces friction and there is a small controlled earthquake.

Continued drilling along the length of a fault, like the San Andreas, would be very expensive. There is also the danger of interfering with unknown forces and doing more harm than good.

Exercise 1: Earthquakes

The word square contains at least 29 words connected with earthquakes. They are printed backwards and forwards, horizontally, vertically and diagonally. Find as many as you can, the list below will help you.

epicentre earthquake tsunamis shake locked fault laser geodolite fall dead kill core damage tremor wave solid shock tragic focus break source movement rebuild Alaska Chile Peru strain repeat release

V	T	A	D	E	K	A	U	Q	H	T	R	A	E
B	S	T	R	A	I	N	U	Q	U	A	K	X	P
O	E	S	A	E	L	E	R	U	C	L	D	O	I
R	E	B	U	I	L	D	N	R	M	E	R	O	C
T	F	G	R	X	G	A	D	E	T	R	T	I	E
R	A	M	W	E	M	D	S	P	O	L	G	D	N
E	V	A	W	I	A	E	E	E	Y	A	L	Y	T
M	G	L	S	H	A	K	E	A	R	S	H	O	R
O	E	A	B	I	T	C	F	T	D	F	O	S	E
R	K	S	M	G	E	O	D	O	L	I	T	E	D
H	C	K	F	A	U	L	T	A	C	H	I	L	E
K	O	A	A	F	D	S	U	M	T	U	K	M	J
T	H	U	L	A	I	E	C	R	U	O	S	X	I
Z	S	O	L	I	D	T	N	E	M	E	V	O	M

(Solution on page 142)

Exercise 2

Write out and complete the following statements using the words listed below.

1 In the Erzurum earthquake more than ________ towns and villages were destroyed.
2 There are ________ severe earthquakes in the world every year.
3 Earthquakes can be ________ by the amount of damage.
4 The Turkish earthquake was ________ on the Richter scale.
5 The ground vibrates because of the sudden ________ of energy.
6 The nearest surface location to an earthquake source is called the ________.
7 Seismometers are ________ used to measure shock waves.
8 L waves cause the most ________ in an earthquake.
9 People were killed at Yungay, Peru because an earthquake set off an ________.
10 Tsunamis were produced in the 1964 ________ earthquake.
11 Designing stronger houses can ________ many lives in an earthquake.
12 Sensitive instruments can measure the build up of ________ in the rocks.

Alaskan epicentre 50 20 instruments 7.6 avalanche release strain measured damage save

Exercise 3

1 Look at Fig 5.1 again.
a) How many plates meet along the Anatolian fault?
b) Explain why earthquakes happen so often along this fault.

2 Explain the terms:
   a) epicentre;
   b) focus.
3 Describe fully the two main earthquake scales and state how they may be used to measure earthquakes.
4 What is a seismograph and how is it recorded?
5 Describe how seismic waves are transmitted through rock. Illustrate your answer.
6 Explain how seismic waves are used to reveal the Earth's interior.
7 a) What is a tsunami?
   b) Why is the wave height of a tsunami so much less in the open ocean compared to near the coastline?
   c) Study Fig 5.11.
      (i) Why are the wave fronts slowed down along the coastlines?
      (ii) The Pacific Ocean measures 15 000 km from north to south. Work out the velocity of the advancing wave in km per hour.
8 a) How long is the San Andreas Fault? (Refer to Fig 5.13.)
   b) What is the average amount of movement on the faultline?
   c) Explain the following:
      (i) a locked section;
      (ii) an active section.
9 Explain how a geodolite could help to predict a possible earthquake.
10 Describe two possible precautions that could be taken to save lives in an earthquake.
11 Read this passage and then answer the question below.

    In the Italian earthquake in 1980 there were at least five violent tremors, in the space of a few hours, in southern Italy. The epicentre of one of these earthquakes was only 10 km from the village of Balvano. People were awakened by the violent shaking of the ground. Within a few seconds most of the houses were reduced to piles of rubble. Many people were instantly buried and the loss of life in one small village was over 100 people.

   What do you think the intensity of this earthquake was on the Mercalli scale? (Use Table 1.)

6 VOLCANOES

Fig 6.1 The Shishaldin Volcano, Aleutian Islands, Alaska, USA

A typical volcano is a cone-shaped mountain (Fig 6.1). It is produced by molten rock called magma reaching the surface. The mountain is formed out of layers of material that have been forced out of a *vent* (hole) in an *eruption*.

ERUPTIONS

A volcanic eruption can last for an hour or for several months, it rarely lasts more than a year. The *active* volcano will then become *dormant* (inactive) until it erupts again. The length of time a volcano stays dormant can vary from a few days to several thousand years. The Helgafell volcano on the island of Heimaey near Iceland, suddenly started to erupt again in 1973. It had been dormant for 7000 years.

If a volcano permanently ceases all activity it is said to be *extinct*. There are no active volcanoes in Britain but there are many extinct ones. For example, Arthur's Seat in Edinburgh, was an active volcano 325 million years ago. The technique of rock-dating has proved this (see p.86). Today just an eroded remnant of the volcano remains.

Volcanic products

When volcanoes erupt they eject solids, liquids and gases. *Lava* (liquid rock) is formed by magma rising from below the Earth's crust. As it flows out of the vent of the volcano, the temperature is usually well above 700 °C. The lava then cools and turns into a solid rock. A lava-flow either has a ropy twisted texture on its upper cooled surface, or a blocky cindery texture (Fig 6.2). This difference depends upon temperature and composition of the lava (see Table 6.1).

Table 6.1

Ropy lava	*Blocky lava*
High temperature (average 1000 °C)	Low temperature (average 700 °C)
Low silica content	High silica content
Flows quickly and easily, often forming a river of lava	Viscous in nature and flows very slowly, often advancing only a few feet per day
Example: Kilauea, Hawaii	Example: Mt Etna, Sicily

Pillow lava may also form if the eruption is below sea level (see Chapter 2, page 12).

Ash

Ash is made of fine powdered rock and lava droplets. It is formed in violent explosions within the volcano. Gases trapped by the rock above them, expand rapidly in an explosive eruption. The explosion shatters the rock, turning it into powder and tiny lava droplets. This fine dust is forced out of the volcano in a dense hot cloud. Some of the ash may settle on the slopes of the volcano but much of it remains suspended in the atmosphere for many months. The finer ash may be carried by the wind for thousands of miles before it

Fig 6.2 Part of ropy lava formations on top of blocky lava, James Island, Galapagos

finally settles. Some scientists believe large ash eruptions could affect the climate by blocking out the sun's heat.

A layer of compacted ash is known as volcanic *tuff*. Sometimes lava spray will settle as a tuff deposit and then re-melt. This is known as a *welded-tuff* or *ignimbrite*.

Bombs

Sometimes a volcano will throw out large blobs of molten lava called *bombs* or *lapilli*. Lava bombs are ejected with such force that they may travel several kilometres from the vent. As they pass through the air the outer skin cools. When they land, they may break apart to reveal a soft molten centre.

Recently, a party of tourists were climbing towards a volcano in Japan when it started to erupt by throwing out bombs. Although the tourists were several kilometres away, three people were killed as the hot rocks rained down on them.

What causes an eruption?

Many volcanic eruptions are very violent because of gases held at high pressure within the magma. The magma slowly rises towards a blocked vent. Sometimes this can cause the whole mountain to bulge upwards and outwards just before an eruption. Then the gases expand and clear the vent, forcing out ash and lava. This bulging occurred on Mt St Helens in the USA, just before it erupted in 1980.

When the cap of a shaken lemonade bottle is suddenly removed the dissolved gases in the liquid expand. The fizzy liquid foams out of the bottle. Gases in a magma chamber behave in much the same way. They help to drive out the molten lava. Very frothy lava can turn into a solid rock called *pumice*. This is so full of *vesicles* that it will actually float in water. It looks very much like a piece of 'Aero' chocolate. Vesicles form when the gas bubbles become trapped as the rock cools.

Water vapour and carbon dioxide are the most common gases in volcanic eruptions. Water may enter the volcano along the cracks from a nearby ocean. On contact with the magma it instantly vapourises.

During an eruption, the cloud of ash and gases above the cone gives the impression that the volcano is 'burning and smoking'. In reality there is very little real combustion going on.

Krakatoa

In the eruption of Krakatoa, near Java, in 1883, massive explosions ripped apart a volcanic island. Almost the whole volcano disintegrated. The noise was heard nearly 5000 km away. Huge quantities of ash were

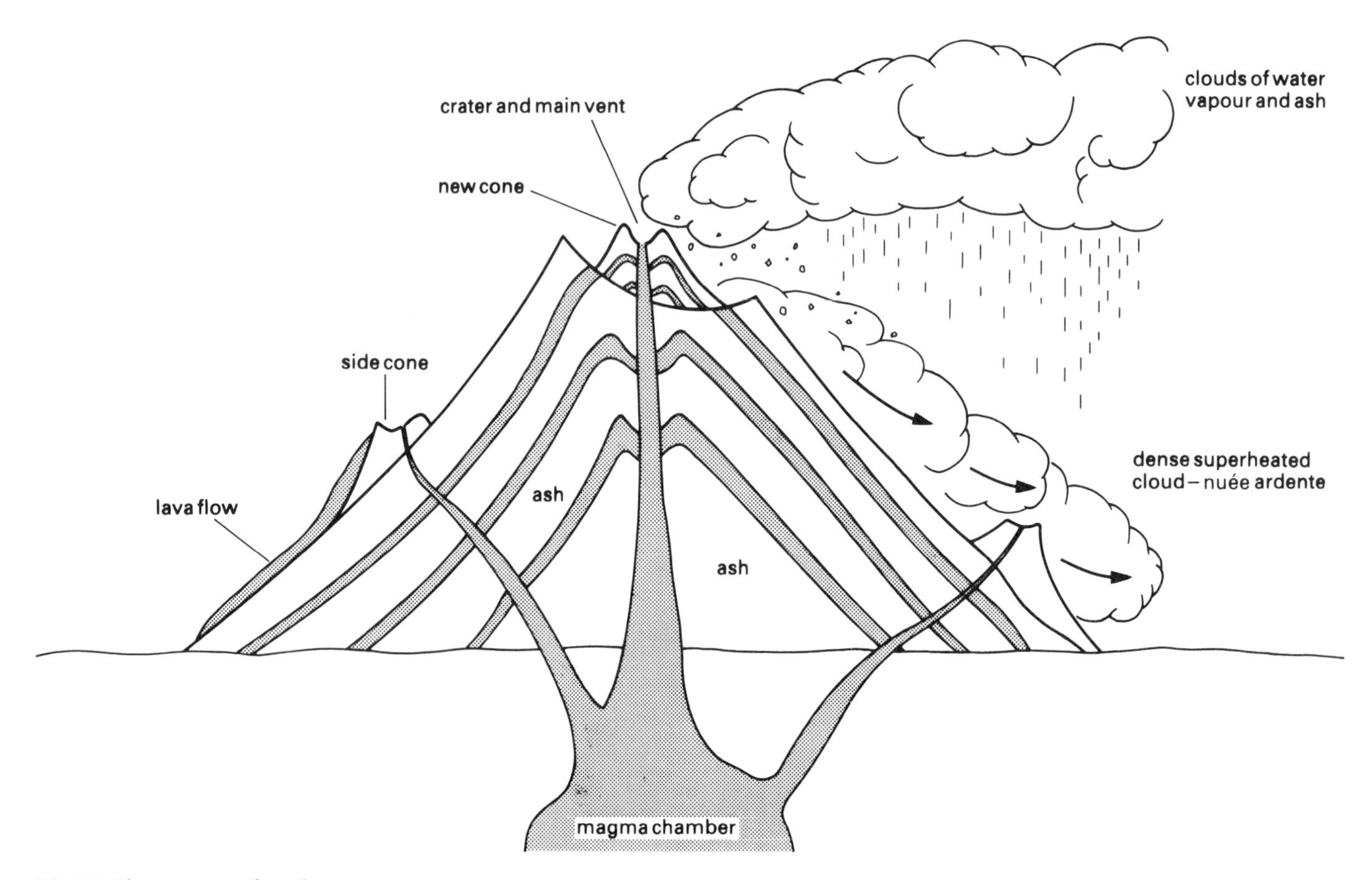

Fig 6.3 The structure of a volcano

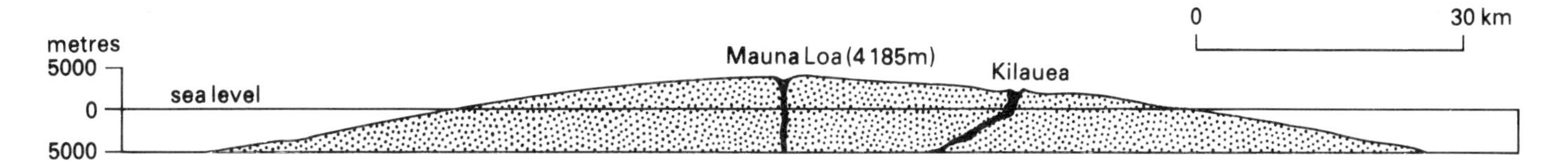

Fig 6.4 The shield volcano that rises above sea level to form the island of Hawaii

blasted more than 80 km into the sky, causing a lowering of temperatures throughout the world as it screened out the sun's heat. One centimetre of ash settled out as far away as Spain. The shock waves caused a huge tsunami that drowned nearly 40 000 people living on coastlines near the volcano.

THE STRUCTURE OF VOLCANOES

Many cone-shaped volcanoes have a *composite cone* made of layered ash and lava. Ash is produced as the vent is cleared at the start of an eruption when gas pressures are at their greatest. This early *explosive phase* is followed by the pouring out of lava in a more *settled phase* as the gas pressure lessens. The erupted lava flows down the slope on top of the ash flow. When the volcano erupts again another layer of ash is produced, followed by more lava. In this way the cone is built up (Fig 6.3).

Variations in shape

The shape of the cone depends on the nature of the erupted material. Volcanoes that are explosive tend to produce steep-sided cones of ash and cinders, and lava flows are of the blocky type, and slow moving. The lava builds up near the vent and can even block it, causing a build up of gas pressure and consequently more explosions.

Sometimes a dense superheated cloud of gas and dust escapes from a crack in the cone below such a blockage. It rolls down the mountainside as a fiery avalanche, travelling at over 100 km an hour and destroying everything in its path. This is known as *nuée ardente* (French for 'glowing cloud') (Fig 6.3). The town of St Pierre in Martinique was completely wiped out in 1906 when Mt Pelee erupted in this way. The nuée ardente smothered the town and 30 000 people were killed.

At the other extreme are lava volcanoes which erupt with very little explosive activity. The lava is runny and flows a long distance from the vent. A large shallow sloping cone is built up to form a *shield* volcano, eg, Mauna Loa on Hawaii (Fig 6.4).

DISTRIBUTION

There are about 500 active volcanoes in the world today. Approximately 20 volcanoes will erupt in any one year. Their distribution is shown in Fig 6.5. You will see that volcanoes are usually found at plate

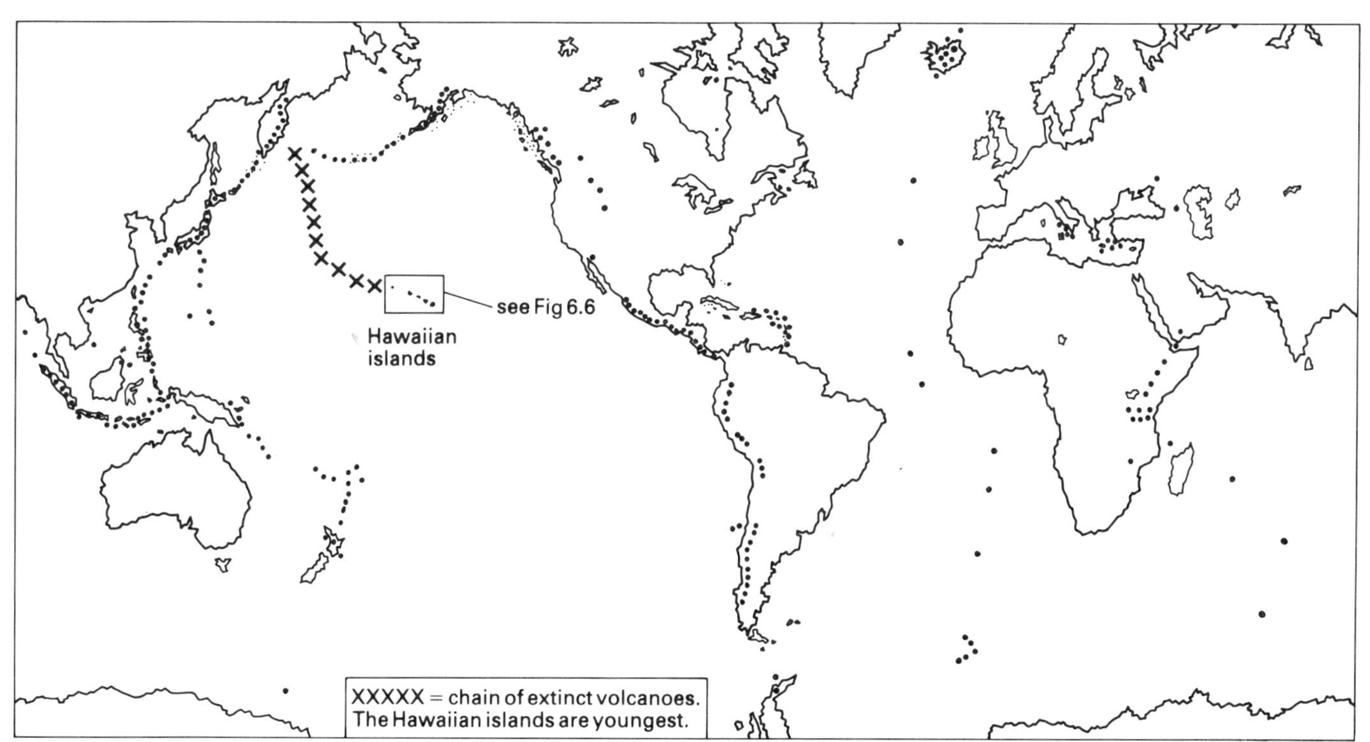

Fig 6.5 Distribution of historically-active volcanoes

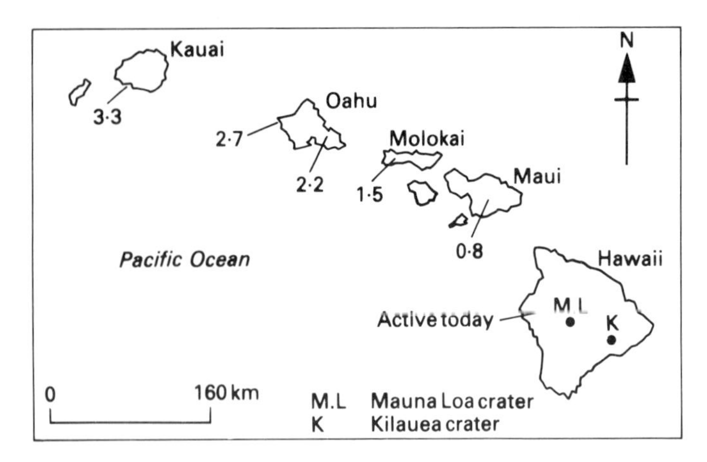

Fig 6.6 The Hawaiian islands. The figures give the age of the lava in millions of years. The younger volcanoes form the larger island. Older volcanoes in the chain are submarine seamounts.

Activity 2: Making a volcano

Warning The green powder (chromic oxide) is a harmful irritant. This activity should only be demonstrated by your teacher using a fume cupboard.

First make a papier mâché model of a cone-shaped volcano. Place a 10p piece in the crater and cover it with one teaspoon of ammonium dichromate powder. Light using a bunsen burner, then close the fume cupboard door.

The expanding chromic oxide which is formed represents the ash. It is a good idea to see this effect once in a darkened room. Then repeat the experiment in a well lit room in order to see the layering of ash on the cone more clearly.

Note how most material collects near to the vent, layering the slopes. Sketch and describe your findings. What type of volcanic eruption does this model best represent?

---

boundaries. They appear along the mid-ocean ridges, as well as along the collision edges of plates. They are especially common around the Pacific. This line is sometimes called 'ring of fire'. Very few volcanoes are found far inland.

WHAT CAUSES VOLCANOES?

The simple answer to this question is to say that they are caused by rising magma – but why does magma form in the first place?

At one time it was believed that there was a layer of magma everywhere below the crust, but the behaviour of seismic waves (see page 29) shows this to be false. In fact, magma exists in isolated pockets. These pockets are especially common at ocean ridges, where there is tension. Rising convection currents in the mantle ensure that the rocks are hot enough to melt. Where there is release of pressure along tensional faults, the magma can get to the surface.

Volcanoes are also found at ocean trench margins of plates, where one plate is undergoing subduction (see page 23). As the plate sinks, frictional heat can create magma. Also, the top surface of the plate has water within its structure. The presence of water lowers the melting point of rocks. Magma reaches the surface in bubbles known as plutons.

Hawaii: a case study

Rarely, volcanoes are found in the centre of a plate. Some of the best examples are the volcanoes of the Hawaiian Islands. Hawaii is the main island at the end of a long chain of submarine volcanoes which stretch to the Aleutian Ocean trench, over 6000 km away. All the islands are made of *basalt* lavas. Dating of these lavas shows that the oldest volcanoes are to the north-west and the youngest to the south-east. The two main active volcanoes, Mauna Loa and Kilauea, are on Hawaii itself (Fig 6.6). They are the largest volcanoes on Earth, standing up to 10 000 m above the ocean floor.

The explanation for the volcanoes' existence in the centre of a plate, is that underneath the Pacific plate is a stationary 'hot spot' in the mantle. As the Pacific plate moves to the north-west, magma from the hot spot burns its way to the surface. Over time a successive chain of volcanoes has been formed in this way.

This effect can easily be seen using a polystyrene sheet-tile to represent the plate, and a soldering iron to represent the stationary magma body. As the tile is passed slowly over the hot soldering iron, it melts a hole through to the surface. This creates a linear pattern of burns on the tile much like the Hawaiian chain of volcanoes.

Calderas

The crater at the summit of Mauna Loa is five kilometres long and two and a half kilometres wide. Large craters like this are known as *calderas*. A caldera is a *collapse structure*, the top of the mountain collapsed into the space left in the magma chamber after an eruption. A typical eruption at Mauna Loa begins by the caldera filling up with lava. A lava lake forms which eventually overflows down the mountain-side (Fig 6.7).

EXPLOSIVE VOLCANOES

In contrast to eruptions of a Hawaiian type are the explosive volcanoes, which are far more dangerous. Many explosive eruptions are of massive proportions and are capable of destroying all life surrounding the volcano. In the 1980 Mt St Helen's eruption in Washington, USA, 60 people were killed as a sideways blast of superheated ash roared out of the mountain. Whole forests up to 30 km away were flattened like matchsticks into the hillsides.

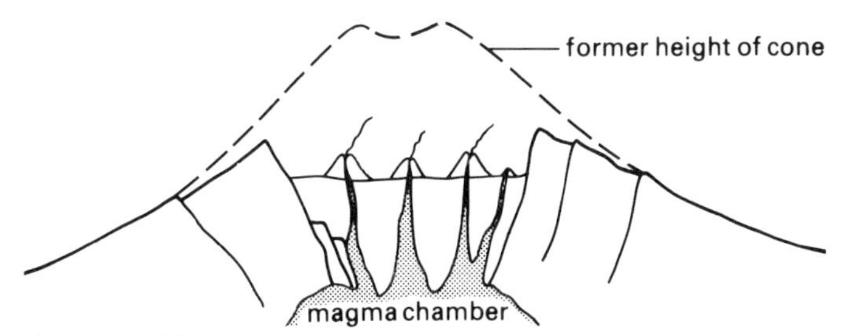

Fig 6.7 A caldera. New cones develop within the caldera crater that has formed either by collapse into the magma chamber, or by explosive activity.

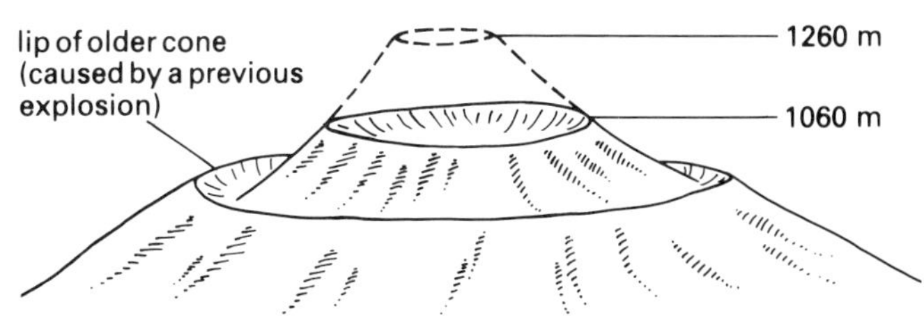

Fig 6.8 El Chichon, Mexico. The top part of the cone was blown away, together with many tonnes of underground magma.

El Chichon: a case study

El Chichon is a volcano in a remote part of southern Mexico (Fig 6.8, and see page 38). The soil on the slope of the volcano is especially fertile because of the dissolved minerals from the ancient ash deposits. At the beginning of 1982, there were seven towns and villages with over 40 000 inhabitants clustered on the lower slopes of the mountain. These people were happy to live there because good crops grew in the rich soil, and El Chichon had never erupted in living memory.

It was true that every so often people could hear rumblings from the mountain, and there was always a cloud of hot water vapour at the summit, but everyone was used to these things. In fact the volcano had remained dormant for over 1000 years. The last major eruption was 15 000 years ago. Some people started to worry when there were over 500 small earthquakes during March. Then suddenly on Sunday, 28 March 1982, a column of rocks, ash and gases shot up 17 km into the sky. A few people were killed, mainly by falling rocks. Many more did not know it, but they had less than a week to live.

Following the eruption, the authorities decided to evacuate the villages but the people were reluctant to leave their land and their homes. They said the volcano had gone quiet again. Once the army had left many people returned to their homes, ignoring the notices saying **PELIGRO: VOLCAN** (Danger: Volcano).

Francisco Leon

Francisco Leon was a village of 1000 inhabitants, only six kilometres from the mountain. After the first eruption many people left the village but about forty stayed on, together with the scientists who were keeping a close watch on the volcano.

On Saturday, 3 April, the mountain exploded. For two days the volcano roared, spewing out thousands of

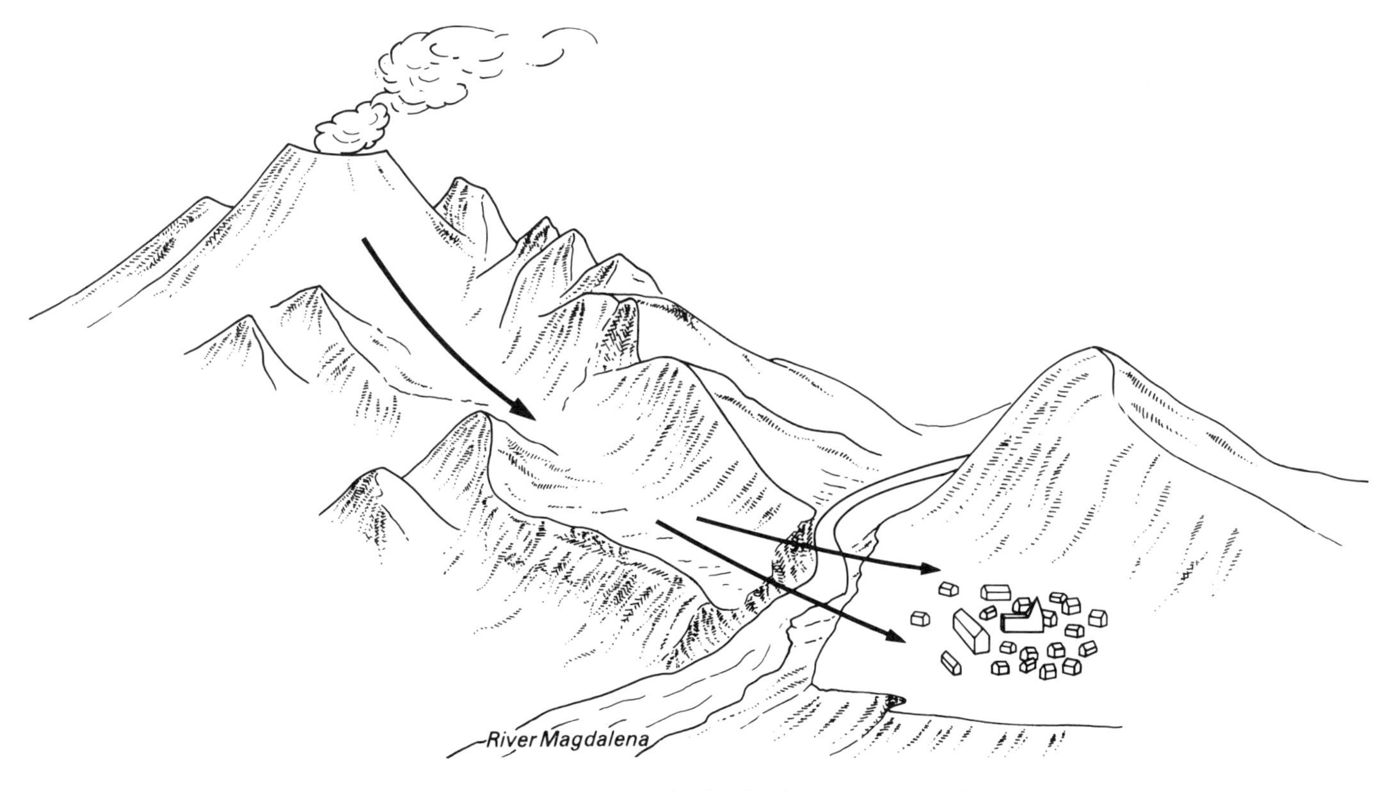

Fig 6.9 The destruction of Francisco Leon. The arrows show the path taken by the glowing avalanche.

tonnes of ash, rocks and gases. The only light during the time was from frequent discharges of lightning in the towering ash clouds. No one from Francisco Leon was seen alive again. When the village was finally reached two weeks later the only thing visible above a thick covering of ash was part of the church. One of the first bodies found was that of a boy holding on to his puppy.

Francisco Leon was destroyed by a nuée ardente. Imagine a dense cloud of fast-moving hot gases, ash, rocks and lava droplets rolling down the mountainside at 100 km per hour. As it approached the village the glowing avalanche was channeled between two hills (Fig 6.9). It then 'ski-jumped' the river and poured into the village on the opposite bank. The temperature of the cloud was estimated to have been between 300–350 °C when it reached Francisco Leon. The Italian city of Pompeii was destroyed and buried in a very similar way when Vesuvius erupted. The people who did not burn to death, choked and suffocated in the poisonous gases.

The effect on the global climate

CHILLY?

Blame it on El Chichon

5 April 1983

A university weather expert said yesterday "It snowed through the Easter weekend and this summer will be poor – a time for brollies and woollies and not much sunbathing. The summer following a volcanic eruption is usually poor. If you are planning a warm holiday this year you need to go a long way south, as far as the Sahara."

Most of the material from El Chichon was blasted vertically into the atmosphere. Some of the 500 million tonnes of material settled out fairly quickly.

Fig 6.10 The Heimaey eruption, showing houses buried in ash. Over 100 houses were buried in the first two weeks of the eruption.

However, much of the finer dust and sulphur dioxide entered the upper atmosphere to a height of 38 kms where it spread as a huge cloud round the globe. The cloud cut down the amount of sunlight reaching the Earth for over two years. This made the Earth slightly cooler and could have been responsible for major changes in the weather. Predicting the effects on the weather are difficult. So many other things can affect the weather pattern. This difficulty is clearly shown by the newspaper account.

As it turned out, Britain had one of its hottest and driest summers on record, with temperatures in late June and July soaring to over 30 °C! In contrast, Iceland had one of its coolest and wettest summers.

FORECASTING ERUPTIONS

The behaviour of many volcanoes is very difficult to predict, especially the more explosive types. Different monitoring systems are used by scientists to try and predict eruptions. For example, the Hawaiian volcanoes swell slightly so sensitive *tiltmeters* are used to monitor them. These are elaborate spirit-level type instruments that record the changes in slope angle. The main difficulty for scientists is saying exactly *when* a volcano will erupt. The very slow swelling of the summit can be as little as one metre in 10 years!

As the new magma forces its way upwards, small earthquakes are caused. These can be recorded by seismographs. The main clue to an imminent eruption is the number of small earthquakes caused by the rising magma. When an eruption is about to happen, there could be as many as 1000 minor, shallow earthquakes a day. Explosive eruptions are very difficult to forecast and dangerous to study at close quarters. Little is known about the behaviour of volcanoes prior to an explosive eruption.

Once an eruption begins there is very little that can be done to stop it. When an advancing lava flow threatens a town it is possible to try and divert it by bombing. The intention is to bomb the land just in front of the lava flow. This creates a new channel for the lava and diverts it from the town it is approaching. This method has been used with limited success on Mt Etna, in Sicily.

During the eruption of Heimaey off the coast of Iceland in 1973, huge quantities of cindery ash were blown onto the roofs of buildings and lay drifting in the wind like black snow (Fig 6.10). The weight of ash on house roofs was so heavy that many roofs collapsed. For four months, the islanders swept the ash off their roofs and bulldozed it from their streets. They also sprayed millions of gallons of seawater onto the lava flow as it approached the town. This slowed the lava down but did not stop it entirely. Then the eruption ceased. The people now live on a bigger island with a better harbour, and they have used the cindery ash to make a new runway for the airfield (Fig 6.11). The islanders have a problem with salt in their soil. Can you think why?

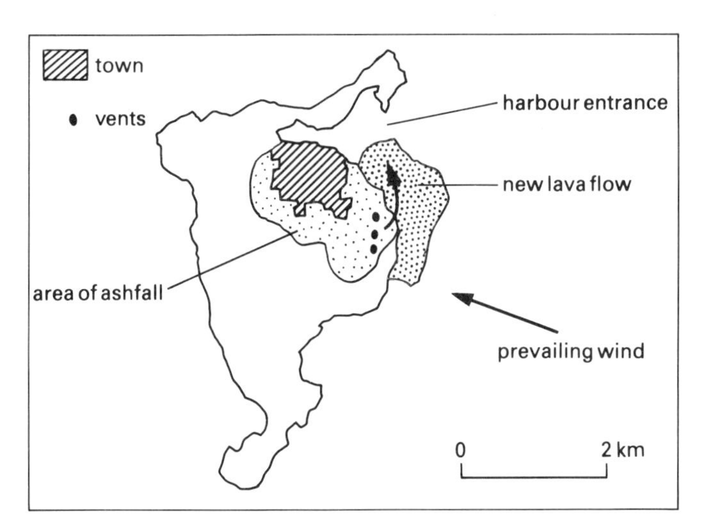

Fig 6.11 The island of Heimaey, south of Iceland

THE EFFECTS OF UNDERGROUND MAGMA

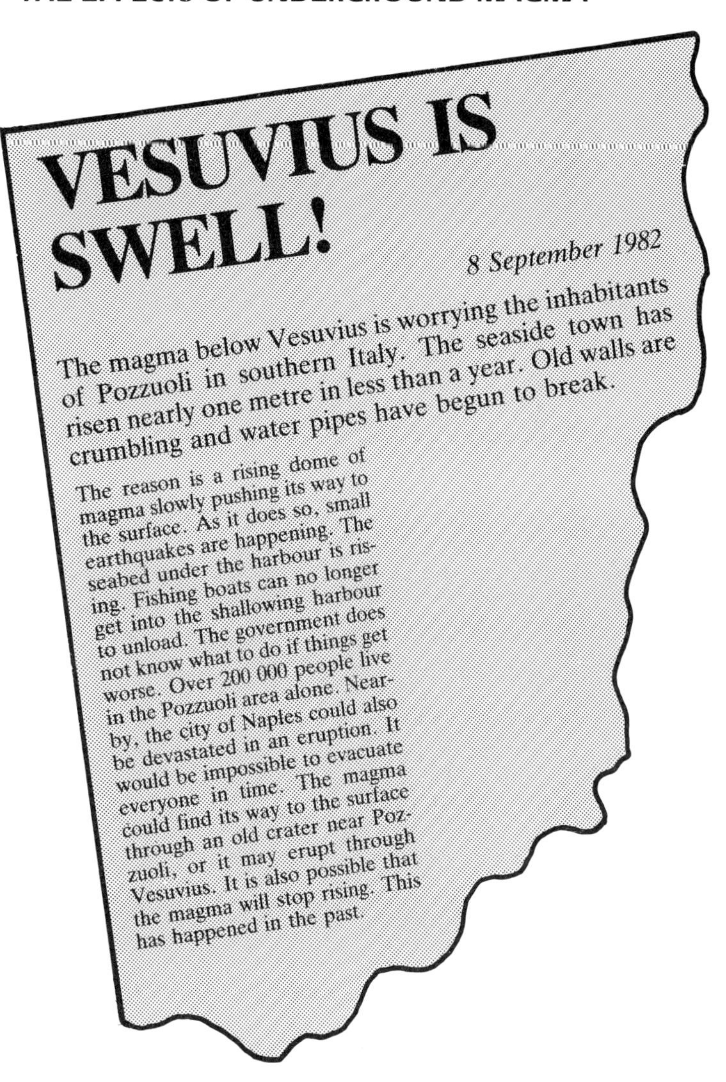

VESUVIUS IS SWELL!

8 September 1982

The magma below Vesuvius is worrying the inhabitants of Pozzuoli in southern Italy. The seaside town has risen nearly one metre in less than a year. Old walls are crumbling and water pipes have begun to break.

The reason is a rising dome of magma slowly pushing its way to the surface. As it does so, small earthquakes are happening. The seabed under the harbour is rising. Fishing boats can no longer get into the shallowing harbour to unload. The government does not know what to do if things get worse. Over 200 000 people live in the Pozzuoli area alone. Nearby, the city of Naples could also be devastated in an eruption. It would be impossible to evacuate everyone in time. The magma could find its way to the surface through an old crater near Pozzuoli, or it may erupt through Vesuvius. It is also possible that the magma will stop rising. This has happened in the past.

INTRUSIONS

Sometimes magma cools and hardens below the surface. Underground rock masses that form in this way are called *intrusions*.

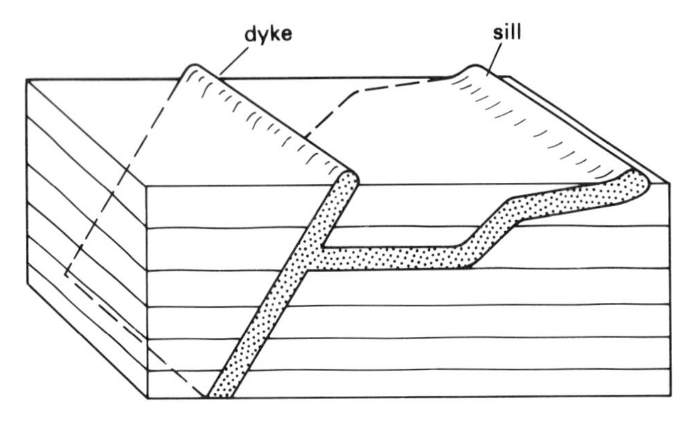

Fig 6.12 Dykes and sills form well below the surface. They are exposed by later erosion, as the land surface is lowered.

Sheet intrusions: dykes and sills

A *dyke* is a thin wall of rock that has cooled and solidified inside a fissure. It cuts across other rock layers (Fig 6.12). A *sill* is formed by a thin layer of magma that has pushed its way between the bedded rock layers, before cooling to form a solid sheet of rock (Fig 6.12).

As the land surface is worn away, dykes and sills often resist erosion and stand out as ridges in the landscape. The Great Whin Sill runs as a near-continuous ridge across northern England from Carlisle to Newcastle. The Romans built Hadrian's Wall along the top of it (Fig 6.13).

Dykes and sills are minor intrusions compared to larger intrusions known as *bathyliths*. A bathylith may be hundreds of kilometres across. It represents the magma chamber and is made of many cubic kilometres of magma. A bathylith is formed by slow cooling of magma deep below the surface (Fig 6.14).

Fig 6.13 The Great Whin Sill and Hadrian's Wall

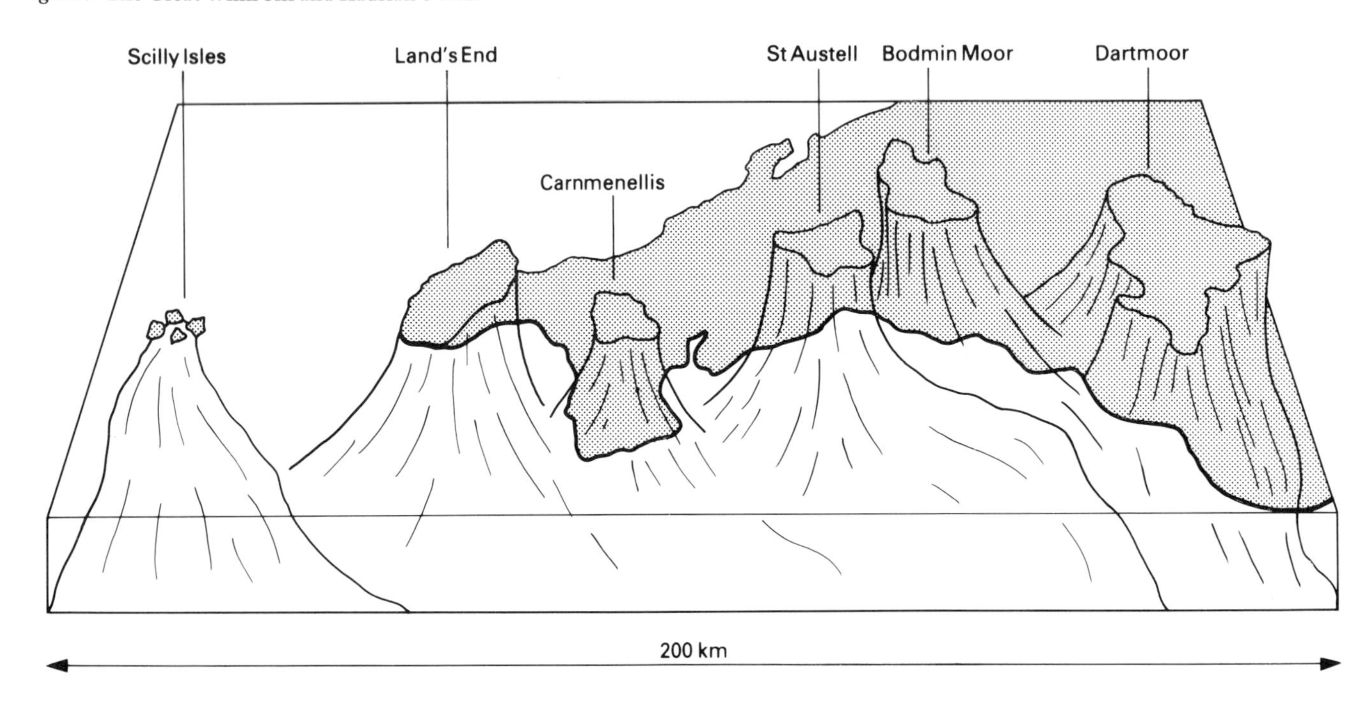

Fig 6.14 The granite bathylith of south-west England. All the granite intrusions are of the same age – 290 million years.

Exercise 1: Volcanoes in action

Find the hidden words. They may be vertical, horizontal, diagonal or even backwards. A list of the hidden words is given below the word square.

(Solution on page 143)

S	U	R	T	S	E	Y	V	V	N	E	T	L	O	M
A	O	H	S	T	H	E	L	E	N	S	L	X	A	S
O	T	N	E	S	B	O	M	B	S	B	E	G	U	T
T	I	U	A	A	O	L	C	F	N	U	M	P	D	E
A	M	A	F	C	T	K	C	K	Y	A	V	F	I	E
K	M	L	B	F	L	A	P	I	L	L	I	I	D	P
A	U	Z	C	X	A	O	Y	A	K	L	E	S	U	T
R	S	U	O	C	S	I	V	C	R	P	N	S	S	S
K	L	L	M	R	A	A	A	E	M	E	O	U	T	L
W	O	T	P	A	B	R	G	O	N	P	I	R	U	Z
O	P	S	O	T	C	G	P	O	T	S	T	E	A	M
L	E	A	S	E	I	E	C	I	M	U	P	N	N	A
L	K	L	I	R	O	C	K	X	L	O	U	L	E	Z
I	Y	B	T	O	E	R	U	S	S	E	R	P	O	V
P	E	L	E	E	V	E	S	I	C	L	E	O	L	Y

volcano ash lava pumice cone eruption steam viscous Vesuvius Pompeii St Helens Pelee fissure Krakatoa Surtsey composite magma shock pillow vesicle crater summit slope steep basalt pipe tuff lapilli rock trigger vent pressure dust crack melt bombs heat pile molten blast

Exercise 2

El Chichon, Mexico

1 a) Copy Fig 6.8. On your sketch, put in a dotted line to show the shape of the older cone.
 b) How high do you think the older cone was?
 c) What happened to this earlier cone?

2 Summarise the main facts about the 1982 El Chichon eruption under these headings.
 a) Warnings and signs of an impending eruption.
 b) The effect of the eruption on the local area.
 c) The possible effect of the eruption on world climate.

Pozzuoli, Italy

3 Explain fully what is happening in the area of Pozzuoli.

4 Explain how the rising magma has already affected the economy of Pozzuoli.

5 What do you think the government should do about the economic problem?

6 Imagine you are a reporter sent to the area because there has been an eruption. Write a dramatic account of what you see when you arrive.

Exercise 3

1 Explain what is meant by the terms:
 a) active
 b) dormant
 c) extinct.

2 Name one example of a British volcano.

3 Draw a labelled diagram of a composite volcano. Then make explanatory notes on all the features shown on your sketch.

4 Trace the world map of volcanoes (Fig 6.5) and overlay the tracing paper onto the map of the world's plates (Fig 2.3).
 a) What do you notice about the location of the majority of today's active volcanoes.
 b) On your overlay, shade in a thick red line to show the 'ring-of-fire' around the Pacific Ocean.

5 What is the main cause of the volcanoes of the Hawaiian Islands?

6 Study the illustration of the Hawaiian Islands (Fig 6.6).
 a) Where is the mantle 'hot spot' today?
 b) Why are the younger islands so much larger than the older ones?

7 Explain what a caldera is and how it is formed. Name two examples of calderas.

8 Explain what happened in the 1980 Mt St Helens' eruption.

9 Describe the role of gases in a volcanic eruption.

10 'Volcanoes that erupt thick pasty lava tend to be explosive in behaviour.'
 a) Explain how variation in lava composition might affect volcanic behaviour.
 b) How are the shape and size of a volcano related to lava composition?

11 Describe the two main ways in which magma can be created at plate boundaries.

12 Explain what is meant by:
 a) a dyke
 b) a sill
 c) a bathylith.

7 MINERALS

ROCKS, MINERALS AND CRYSTALS

The rocks of intrusions and lava flows form from molten magma. Rocks that form by cooling from a molten liquid are known as *igneous rocks*. Igneous rocks are made of *minerals*. Minerals are naturally occurring substances that have their own unique structure, shape and chemical make-up. A true mineral is *inorganic* (non-living), it does not form from the remains of plants or animals. Although more than 2000 minerals are known, nearly all rocks are formed from just a few of them.

Quartz

The mineral quartz is found in many different rocks. Like many minerals it forms attractive *crystals* (Fig 7.1).

Crystals have smooth flat faces and a regular geometrical shape. Liquid magma produces different crystals as it cools. When all the liquid has been used up, the crystals lock together and a solid igneous rock is formed. The pattern of interlocking crystals is known as the *crystal texture* (Fig 7.2).

Fig 7.1 Quartz crystal, showing fine scratches on crystal faces

Fig 7.2 Granite, showing mineral texture

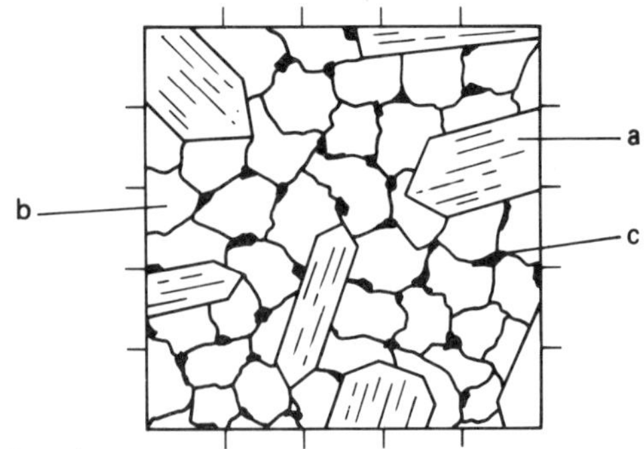

Main minerals

a feldspar – white or pink, pearly-looking

b quartz – grey to white, glassy-looking

c mica – black speckles, flaky, shiny

Fig 7.3 (a) The crystal texture of granite

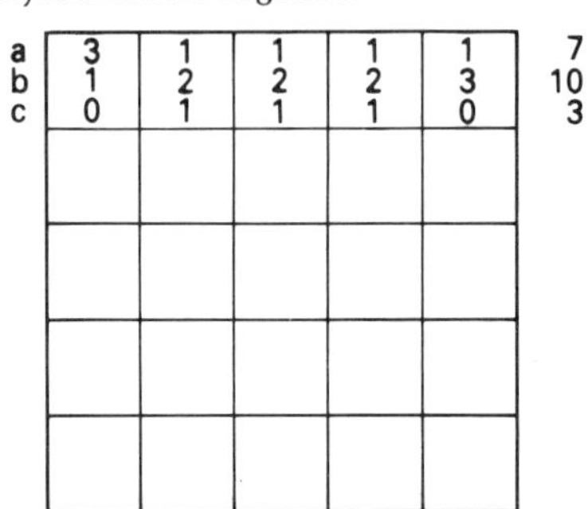

a	3	1	1	1	1	7
b	1	2	2	2	3	10
c	0	1	1	1	0	3

Fig 7.3 (b) The grid (for use with Activity 3)

Activity 1: The crystal texture of granite

1 Study the sketch of the crystal texture of granite (Fig 7.3), and identify the minerals in a specimen of granite.
a) Which mineral has the largest best-formed crystals? Can you think of a reason?
b) Which mineral is present in the greatest proportion?

2 To measure the proportions of each mineral in the granite, copy the grid (Fig 7.3b) onto an acetate sheet. As the 25 squares in the grid = 100%, one small square = 4%.
Place the grid over Fig 7.3(a), then estimate the mineral content of every square in the grid. This has already been done for the top row of the grid.

3 Finally add up the total for the three minerals to find the percentage of each one in the sample.

4 Use the grid on a real rock sample.

Activity 2: Breaking up granite

Since the minerals in granite are quite large it is possible to separate them out.

NB Wear safety glasses during this activity.

1 First try to break up a piece of granite, using a hammer. You will find it very difficult. Granite is an extremely hard rock.

2 Heat up the granite on a gauze mat over a bunsen burner, or in a hot kiln, for several minutes.
3 Carefully remove the hot rock with tongs and plunge it into a bath of cold water.
4 Remove the cold rock from the water and attempt to break it up again. This time the rock will fall apart easily.
5 Sort out the three main minerals into separate piles. *Quartz* has a glassy shine and is grey in colour. Two kinds of feldspar are often present: *orthoclase* is pink or pearly white and *plagioclase* is white. *Mica* (biotite) is made of small black flakes (Fig 7.4).
a) Why do you think the rock was so much easier to break up after heating and rapid cooling?
b) What was the percentage of each kind of mineral in the piece? How did the results of the experiment compare with the results obtained from your analysis of Fig 7.3 (a)?

ELEMENTS, COMPOUNDS AND ATOMS

Elements

The minerals found in granite are complex substances. Each mineral is made of more than one *element*. An element is a substance that cannot be chemically broken down any further. Sometimes a mineral is made of just one element. Gold is both an element and a mineral; diamond is made of the single element, carbon. There are about 100 known elements but only eight are very common. They make up 99% of the Earth's crust (see Table 1).

Table 1 Proportions of elements in the Earth's crust

	%		%
Oxygen (O)	47	Calcium (Ca)	3.5
Silicon (Si)	28	Sodium (Na)	3
Aluminium (Al)	8	Potassium (K)	2.5
Iron (Fe)	5	Magnesium (Mg)	2

Compounds

Many minerals are compounds of two or more elements. For instance, the mineral rock salt (Na Cl) is a compound of sodium (Na) and chlorine (Cl). Quartz (SiO_2) is a compound of silicon (Si) and oxygen (O_2).

Atoms

The smallest part of an element that can naturally exist on its own, is called an *atom*. Atoms are extremely small – one grain of sodium chloride (table salt) contains several million atoms. The way atoms are arranged inside the mineral structure is most important; so is the *atomic structure*, the way atoms are linked with each other. The atomic structure decides many properties of the mineral, such as hardness and shape. The sodium (Na) and chlorine (Cl) atoms in salt are arranged in a regular cubic pattern (Fig 7.4). When crystals of salt grow in a liquid, millions of atoms of sodium and chlorine are joined together in this pattern and cubic crystals are formed.

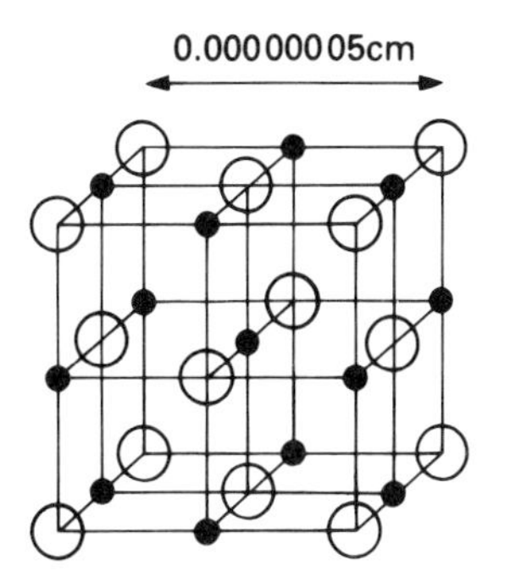

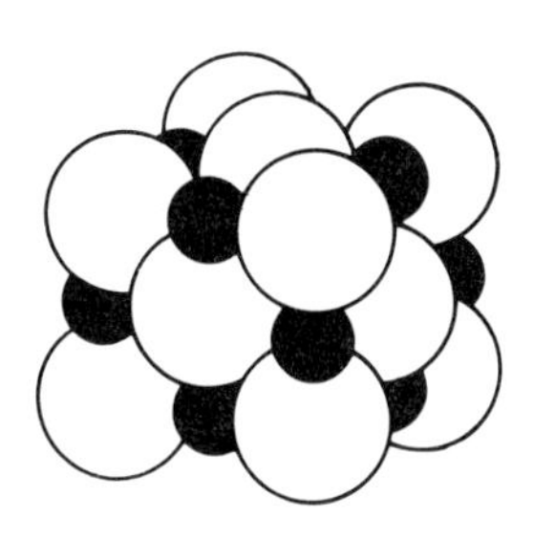

Fig 7.4 The atomic structure of sodium chloride (Na Cl). The larger chlorine atoms pack together, with the smaller sodium atoms filling the spaces between them. When this structure is repeated many times, a visible cubic crystal is produced.

Activity 3: How atomic structure can affect the shape of salt crystals

1 Dissolve some table salt in warm water in a dish. Make sure all the salt dissolves and that the solution is saturated (as salty as possible).
2 Leave the dish in a warm place until all the water has evaporated, and salt crystals have formed. Thoroughly dry the crystals and then examine them. What shape are they?
3 Crush some of the crystals using the flat part of a knife blade. Look at the salt dust with a hand lens. What shape are the tiny crushed grains? Compare them to the salt grains in table salt – are they the same shape?

Conclusion

The atomic structure of the sodium and chlorine atoms controls the shape and the growth of crystals. Crushing salt crystals simply creates small cube fragments because they break along lines of weakness in the atomic structure. These lines of weakness are called *cleavage planes*. They are found in many different minerals (for more details, see page 47).

Diamond and graphite

The arrangement of atoms can also affect the *colour* and *hardness* of a mineral. A good example of the importance of the internal structure is the case of diamond and graphite. They are both made of the same element, carbon, yet they form completely different minerals.

Diamond is the hardest-known mineral whereas graphite is very soft and flaky. Diamond forms clear almost colourless shiny crystals in its pure state;

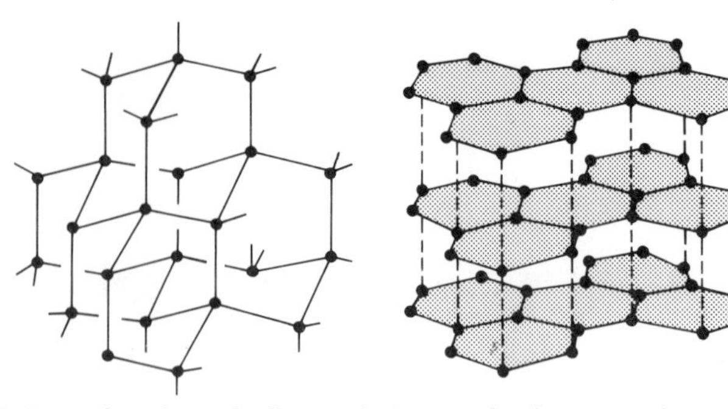

No lines of weakness in diamond, so it is a very hard mineral. Layers of carbon atoms in graphite easily separate to make it a very soft mineral.

Fig 7.5 The structure of diamond and graphite

whereas graphite is dark grey to black and splits easily.

These differences are due to the way the carbon atoms are arranged (Fig 7.5). Diamond forms deep in the Earth's crust under intense pressure. The carbon atoms are packed close together in a locking pattern. In graphite the carbon atoms are arranged more loosely as separate sheets. These sheets easily break apart, this makes the mineral very soft. Pencil lead is made of graphite mixed with clay. When you write with a pencil tiny flakes of carbon break off to mark the paper.

MINERAL CLASSIFICATION

Each mineral has an exact chemical composition. This can be written as a chemical formula. The formula states the amounts of the different elements in the mineral.

The way in which different elements combine can be used to classify minerals. The main mineral groups are shown in Table 2.

Table 2 Minerals grouped according to chemical composition

SILICATES

Quartz; orthoclase feldspar; plagioclase feldspar; muscovite mica; biotite mica; hornblende; augite; olivine.

NON-SILICATES

(i) Single (native) elements
eg gold (Au); silver (Ag); sulphur (S); diamond (C); graphite (C).

(ii) Oxides
eg haematite ($Fe_2 O_3$); magnetite (Fe_3O_4).

(iii) Sulphides
eg galena (PbS); pyrite (FeS_2); sphalerite (ZnS).

(iv) Carbonates
eg calcite ($CaCO_3$); dolomite ($CaMgCO_3$); malachite ($CuCO_3$).

(v) Halides
eg fluorspar (CaF_2); halite (NaCl).

(vi) Sulphates
eg barytes ($BaSO_4$); gypsum ($CaSO_4 + 2H_2O$).

The silicates play a very important role because they are the rock-forming minerals. Over 90% of the Earth's crustal rocks are made of them. All the silicates are built up of atom groups known as *silicon-oxygen tetrahedra*. This basic building unit consists of four large oxygen atoms surrounding a much smaller, central silicon atom. These tetrahedra join together in various ways. This controls the characteristics of the different silicate minerals.

Vein minerals

In addition to their role as rock-formers, many minerals are also found in *veins*. A vein (lode) is a long fissure through which hot watery fluids can pass. Veins are formed by *hydro-thermal* action. Hot watery solutions circulate through the fissures above a heat source, such as a granite intrusion (Fig 7.6). (See also page 124.) Crystals of vein minerals grow outwards from the rock walls on either side of the fissure.

INVESTIGATING MINERALS

Trying to find out about minerals is rather like detective work. You have to look carefully for clues. The tests explained in this section can be used to identify unknown minerals. Collect together these 20 minerals: quartz, orthoclase feldspar, muscovite mica, biotite mica, olivine, hornblende, augite, graphite, haematite, magnetite, galena, pyrite, sphalerite, calcite, dolomite, malachite, fluorspar, halite, barytes, gypsum. It is a good idea to label your mineral specimens with letters. Later, when you have studied each test, you can examine these minerals to find out their properties, and try to identify them.

The hardness test

Hardness is the resistance of a mineral to scratching. In 1812 a German mineralogist F. Mohs arranged a set of 10 'standard' minerals in order of their increasing hardness, with diamond as 10 on the scale. Each mineral is able to scratch or dent any mineral *lower* in the scale. The hardness of a mineral is always the

Activity 4: The hardness test

1 Label the set of 20 unknown minerals with letters.
2 Scratch each mineral with your finger nail, penknife or other object of known hardness (see Table 1 and Fig 7.8). **NB** When using the penknife always scratch *away* from the hand holding the mineral.
3 Use a hand lens to examine the scratch mark. Make sure the scratch is from the mineral being tested.
4 Record the hardness value of each mineral on your copy of the Record Sheet (Fig 7.7).
5 Check your answers against the mineral table on page 51.

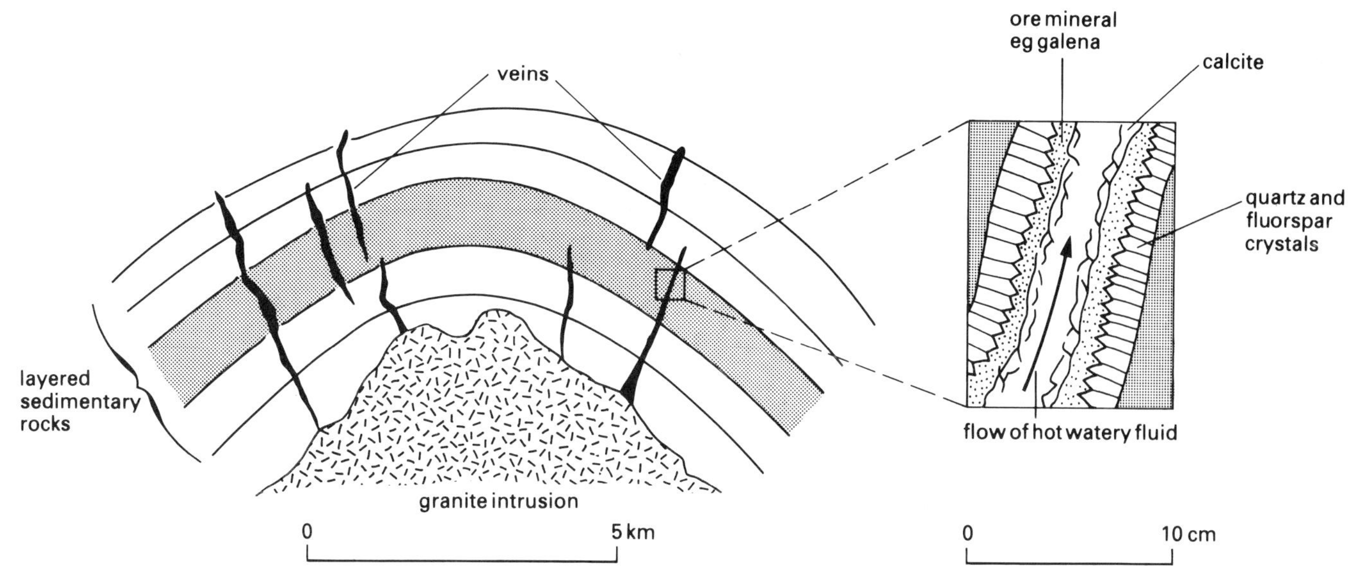

Fig 7.6 **Formation of mineral veins**

same, whatever the variation in colour or shape. Some minerals show great variety of colour, for instance, quartz can be transparent, white, pink, purple, brown or yellow. Whatever its colour, the hardness of quartz is always 7 on Mohs scale. The hardness test is a reliable guide to the identification of an unknown mineral.

Colour

On its own colour does not tell you very much about minerals. They can be many different shades and also, different minerals may have the same colour. However colour is a reliable guide to some of the minerals.

Activity 5: The colour test

Look at the 20 minerals and write down the colour of each one on your Record Sheet.

Table 3 Mohs scale of hardness

List of standard minerals	The hardness values of some familiar items
1 Talc	
2 Gypsum	Finger nail – 2.5
3 Calcite	2p coin – 3
4 Fluorspar	Iron nail – 4
5 Apatite	Glass – 5.5
6 Orthoclase feldspar	Penknife blade – 6
7 Quartz	Steel file – 7
8 Topaz	
9 Corundum	
10 Diamond	

MINERAL	HARDNESS	COLOUR	STREAK	LUSTRE	DENSITY	CLEAVAGE AND FRACTURE	HABIT	OTHER TESTS
a								
b								
c								
d								
e								
f								
g								
h								

Fig 7.7 **The mineral test Record Sheet**

Activity 6: The streak test

1 Rub each mineral in turn against the unglazed back of a bathroom tile, or use a streak plate. The mark made is the streak.
Sometimes the mineral will not mark the tile. This is because either the mineral is harder than the tile or the powder is the same colour as the tile. In this case, scratch the mineral with a penknife blade, being careful to move the blade away from you. Look at the colour of the powder on the scratch.
2 A few minerals are harder than a penknife. Can you work out which minerals they are? To find their streak, scrape each mineral with a steel file. Note the colour of the falling powder.
3 Record the streak colour of each mineral on your Record Sheet and the method used to discover it (ie tile (T), knife (K) or file (F)).

Questions
a) How many minerals have the same streak?
b) What is the most common colour?
c) What is the streak of your finger nail?

Streak

The *streak* of a mineral is a better guide than the external colour. The streak is the colour of the powdered mineral. Whilst the outer colour of a mineral may vary (in different specimens), its streak will always stay the same. Unfortunately, as you will see, quite a few common minerals have a white streak. This means that the test is often not enough on its own, to identify an unknown mineral.

Lustre

Lustre is the extent to which the surface of a mineral reflects light. Some mineral surfaces are very shiny, others are dull. If a mineral looks like a shiny piece of metal, it is said to have a *metallic* lustre. Some metallic minerals oxidise (tarnish) rapidly, and lose their shiny metallic look, eg, rusty iron. Other minerals sparkle in the light and have a *glassy* lustre, eg, quartz. The lustre seen in minerals such as gypsum is softer and less sparkling. Gypsum has a *pearly* or *silky* lustre. If a mineral hardly reflects any light it is said to have an *earthy* or *dull* lustre.

Activity 7: The lustre test

Work through the set of lettered minerals. Write down the lustre of each mineral on your Record Sheet. Use the terms metallic, glassy, pearly silky or dull.

Questions
a) How many of the minerals have a glassy lustre?
b) How many have a metallic lustre?

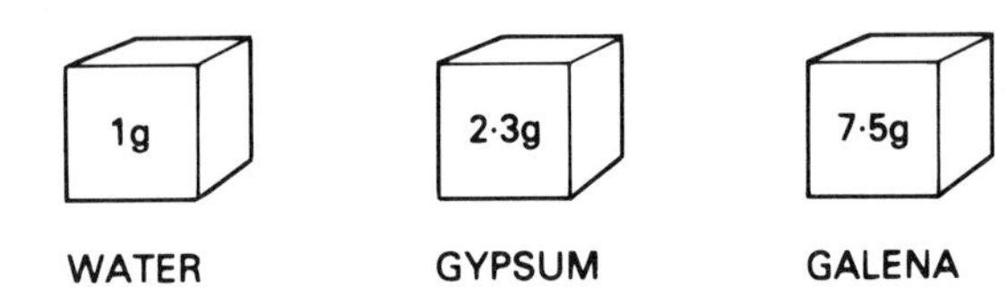

Fig 7.8 The volume of the three substances is the same: 1 cm^3, but the *mass* varies

Density

The *density* of a mineral is its mass in grams (g) packed into each cubic centimetre (cm^3). This is normally written as g/cm^3. Some minerals have high density values, around 6 g/cm^3, especially the metallic ones. A value of around 2.7 g/cm^3 is common for most rock-forming silicates (Fig 7.8).

Measuring density
The density of a mineral can be found by dividing its *mass* by its *volume*. The formula is:

$$\text{Density} = \frac{\text{Mass}}{\text{Volume}}.$$

The *mass* of a mineral is the amount of matter it contains. The *volume* is the amount of space the mineral takes up. The mass can be measured in grams (g) using a top-pan balance.
The volume is measured in cubic centimetres (cm^3) using water in a measuring cylinder. 1 cm^3 of water has a mass of 1 g (=1 ml).
Suppose that a mineral has a mass of 90 g and a volume of 30 cm^3. What is its density?

$$\text{Density} = \frac{\text{Mass}}{\text{Volume}}$$

$$\text{Density} = \frac{90}{30} = \frac{3}{1}$$

So the mineral has a density of 3 g/cm^3.

Activity 8: The density test

1 Weigh each mineral to find out its mass, using a top-pan balance.
2 Then partly fill a measuring cylinder with water. Write down the volume of water.
3 Lower each mineral specimen completely into the water, removing all air bubbles from its surface*. A fine brush is useful for doing this.
4 Record the new water level. The volume of the displaced water is the same as the volume of the mineral.
5 Use the formula:

$$\text{Density} = \frac{\text{Mass}}{\text{Volume}},$$

to work out the density of each mineral.

A Flaky cleavage: in one direction
eg muscovite and biotite mica, graphite and gypsum

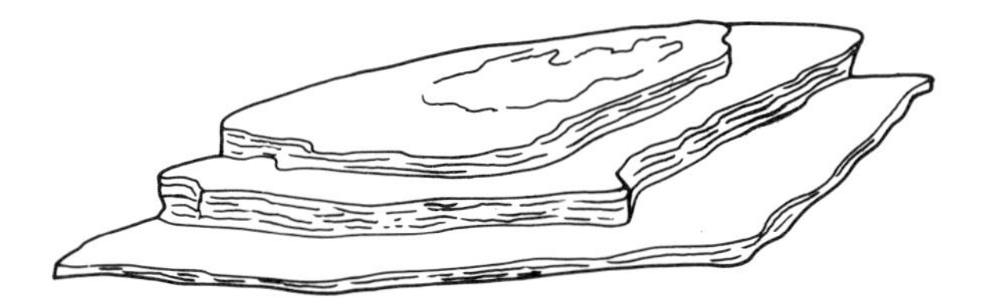

B Rectangular cleavage: 2 planes of cleavage eg orthoclase feldspar

C Cubic cleavage: cleavage planes in 3 directions at right angles
eg galena and halite

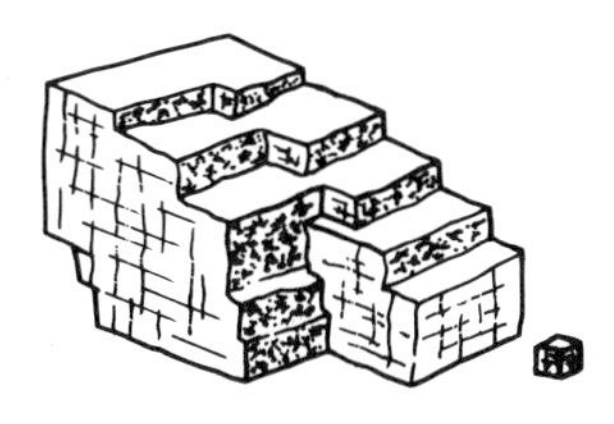

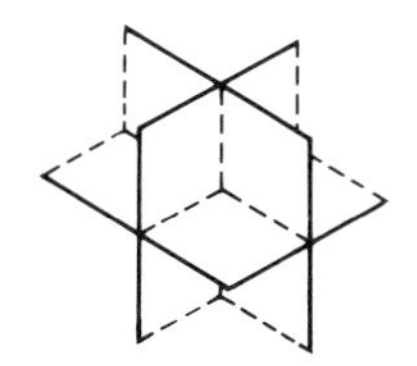

D Rhombic cleavage: cleavage in 3 directions, not at right angles
eg calcite

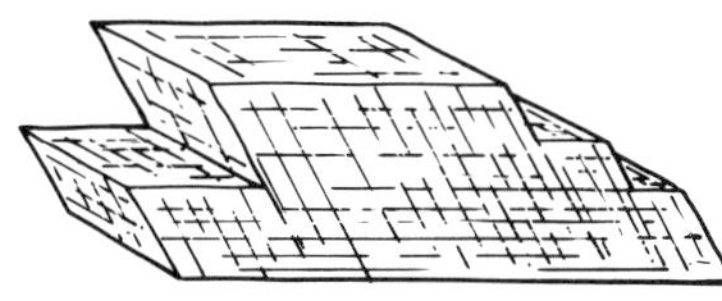

E Octahedral cleavage: in 4 directions

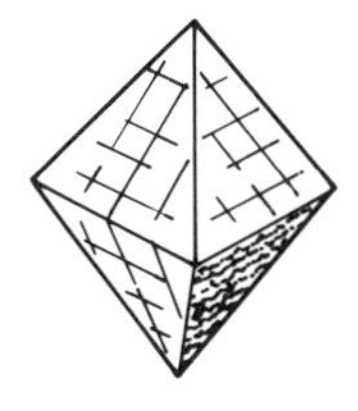

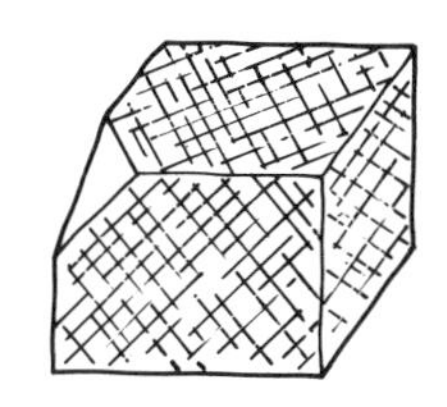

Fluorite crystal with octahedral cleavage

F The difference in cleavage between hornblende and augite

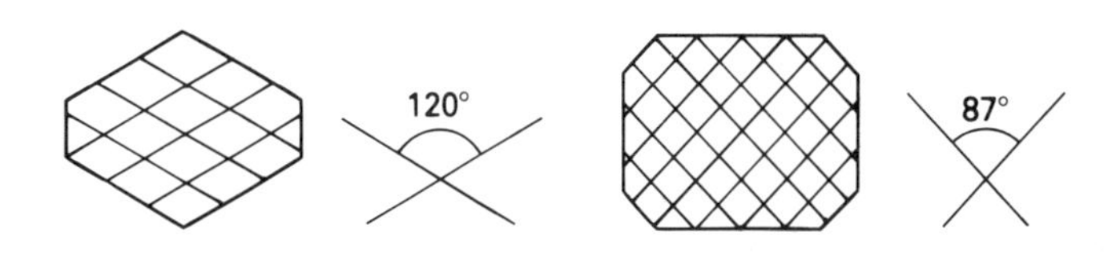

HORNBLENDE: 2 sets at 120°

AUGITE: 2 sets at nearly 90°

Fig 7.9 Cleavage in different minerals

6 Record the densities of the minerals on your Record Sheet.
7 When you have finished check your answers on the Mineral Table on page 51.
8 Underline the values of the three most dense minerals.
* **NB** There is one mineral which dissolves easily in water.

One way to compare densities quickly is to hold the specimen in the palm of your hand. Does it feel heavy or light?

Cleavage

Many minerals easily split up along flat surfaces called *cleavage planes*. Cleavage is caused by weak links between atoms within the mineral structure. This is clearly shown in the case of graphite, see page 44. Some of the main kinds of cleavage are shown in Fig 7.9.

Fracture

Some minerals do not cleave. They will only break up by fracturing. A *fracture* is an uneven break in a mineral which does not follow a cleavage direction. The two main kinds of fracture are shown in Fig 7.10.

Activity 9

Work through the minerals and record the cleavage and fracture properties for each mineral on your Record Sheet. Refer to Figs 7.9 and 7.10 to help you do this.

Where two conchoidal surfaces meet it is possible to make a very sharp cutting edge (Fig 7.11). In many parts of Britain flint is an abundant mineral. It is found as nodules in layers of chalk, and is often eroded out along coastal areas. For example, it forms the major part of the *shingle* beaches in south-east England. Early man relied on flint for making knives, spearheads and arrowheads. Axe-heads and other tools were also made of *obsidian*, a volcanic glass (see Fig 8.2). The secret of manufacture depended on the way flint and obsidian fracture along curved surfaces. If possible, have a look at examples of stone-age tools such as axes and arrowheads in your local museum. Note the sharp cutting edge produced by chipping off the shell-shaped fragments. Imagine how long it must have taken to produce tools like this.

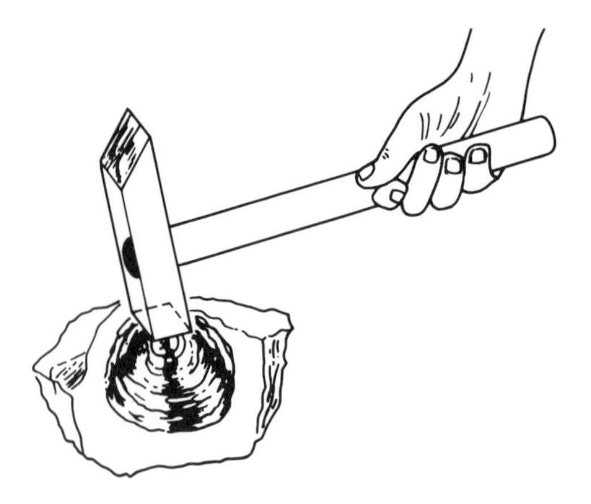

Conchoidal fracture – circular ripples surround the point of impact, producing a shell-like pattern. This is seen in quartz, flint and obsidian.

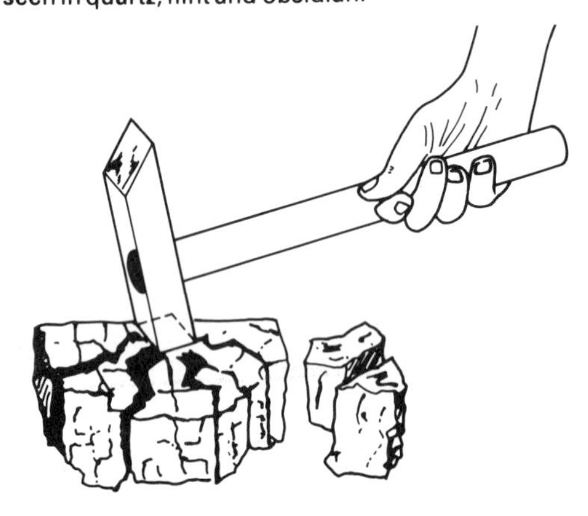

Uneven fracture: the mineral breaks up into rough blocky fragments

Fig 7.10 The two main kinds of fracture

Habit (or form)

When a mineral first forms, it grows by the addition of various elements to its structure. The elements are added in layers to the outside surface of the mineral. Usually minerals grow into a variety of rough shapes because the amount of space in which they can grow is restricted by the other minerals around them. Minerals that are formed in this way are said to have a *massive* habit (form). If a mineral can grow freely – in a vein or cavity for instance, it may form regular shapes with flat surfaces. These shapes are called *crystals*. Many minerals eg halite have a range of different habits, depending on the conditions prevailing at the time of the formation (Fig 7.12).

Some minerals, such as haematite and malachite form special shapes which are not crystals. Haematite usually has a massive or *reniform* (kidney-shaped) habit. It takes the form of rounded kidney-shaped masses. When broken across, the resemblance to a kidney is even more apparent with a fibrous radiating internal pattern of tiny crystals (Fig 7.13).

Fig 7.11 A flint tool. Notice the sharp cutting edge.

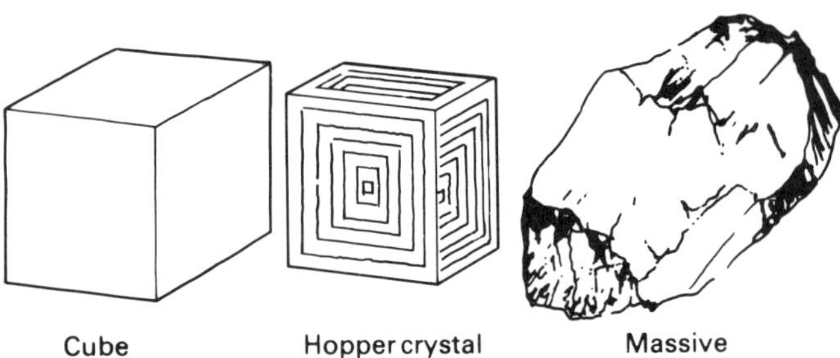

Fig 7.12 The various forms of halite (rocksalt). The photograph shows some of the different forms

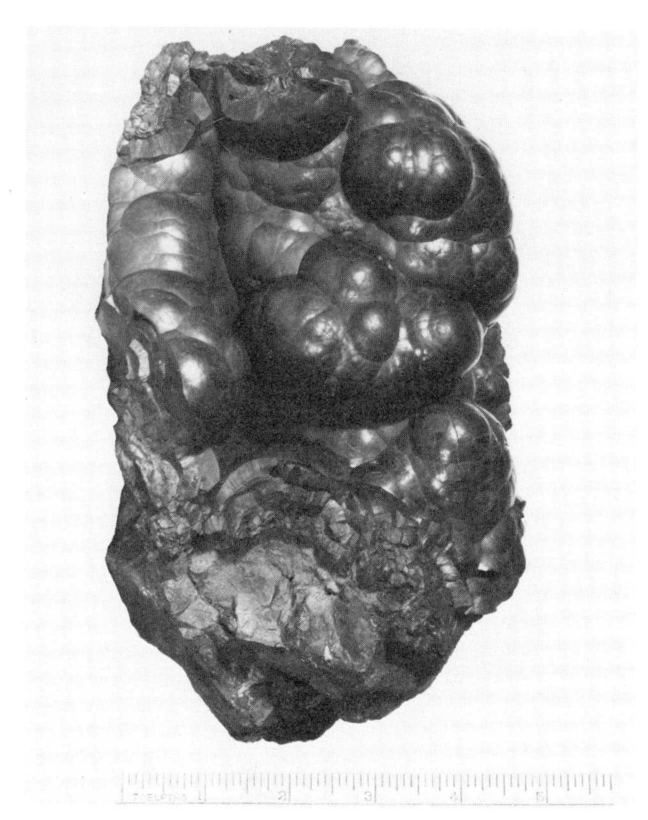

Fig 7.13 The reniform (kidney) shape of haematite

Activity 10: Habit

Examine all the minerals. Write down the habit of each mineral on your Record Sheet, using words like massive, reniform, granular (like sugar grains), fibrous, flaky, cubic crystals, flat crystals, rhombic crystals.

Other mineral tests

Taste

One of the minerals has a salty taste. Can you identify it?

Reaction to acid

When a drop or two of dilute hydrochloric acid is applied to a carbonate mineral, it will fizz. The *effervescence* (fizzing) is the bubbling of carbon dioxide gas. It is released from the carbonate mineral by the acid as the mineral starts to dissolve.

Activity 11: The acid test

Carry out the acid test on each mineral. How many of the minerals react to the acid in this way? Record your findings on your Record Sheet.

A different reaction to dilute hydrochloric acid is shown by one of the metallic minerals. When drops of acid are applied to it, the mineral gives off the pungent bad-egg smell of hydrogen sulphide. Which mineral reacts in this way?

IDENTIFYING MINERALS

By now you should have a full record of the properties of the 20 minerals you have examined. Check your answers against Table 4 (page 51). Discuss any differences between your Record Sheet and the table with your teacher, and correct any errors.

Certain properties are more *diagnostic* than others, in other words, more useful for identification purposes. The more diagnostic properties are marked with the symbol ■ on the mineral table. To help you find out

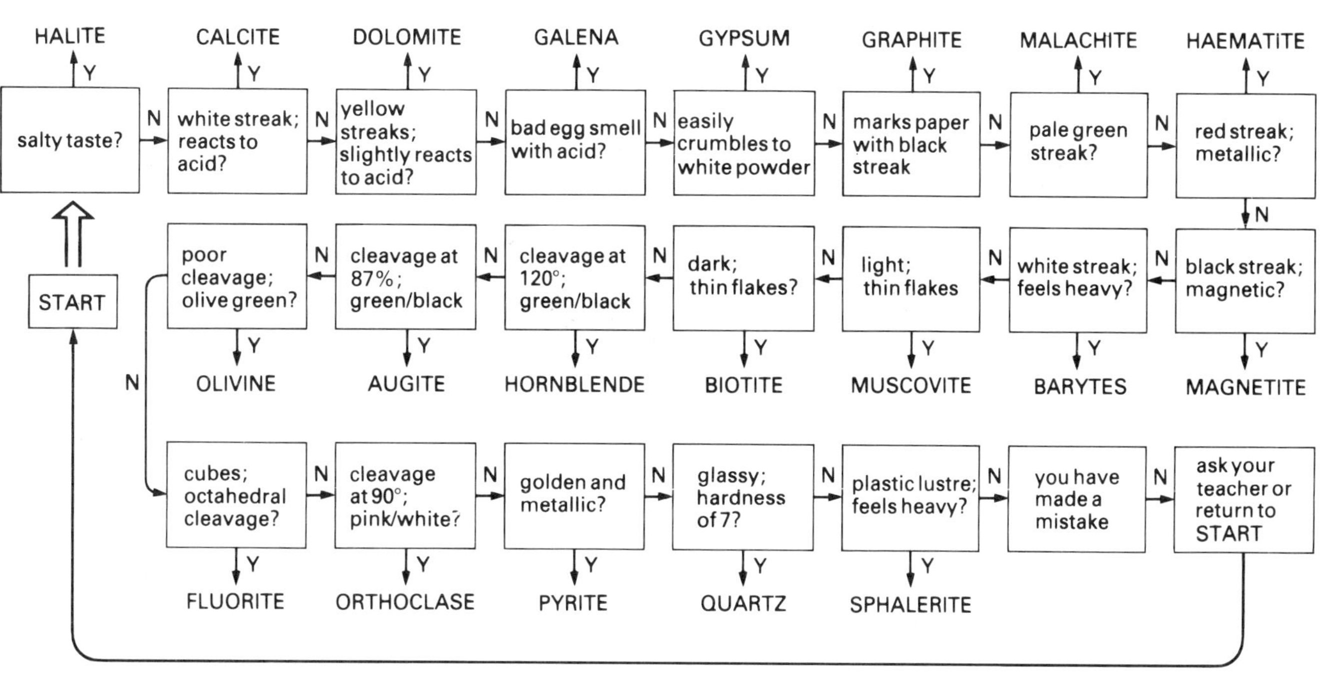

Fig 7.14 Flow diagram for mineral identification

the names of the minerals, work through the flow-diagram in Fig 7.14. This singles out the more diagnostic properties of each mineral. Alternatively, you could use the computer program (see Appendix 1). The program is based on the flow-diagram and will give you the name of each mineral.

Finally, Table 5 provides some more information on the main occurrence and use of each mineral. Some minerals are described as being important *ores*. Ores are valuable minerals that have an industrial use and it is economically viable to extract them. You can read more about the search for ore minerals in Chapter 11.

Exercise 1

Make your own word square puzzle. Use words to do with minerals. Then test the puzzle on your friends!

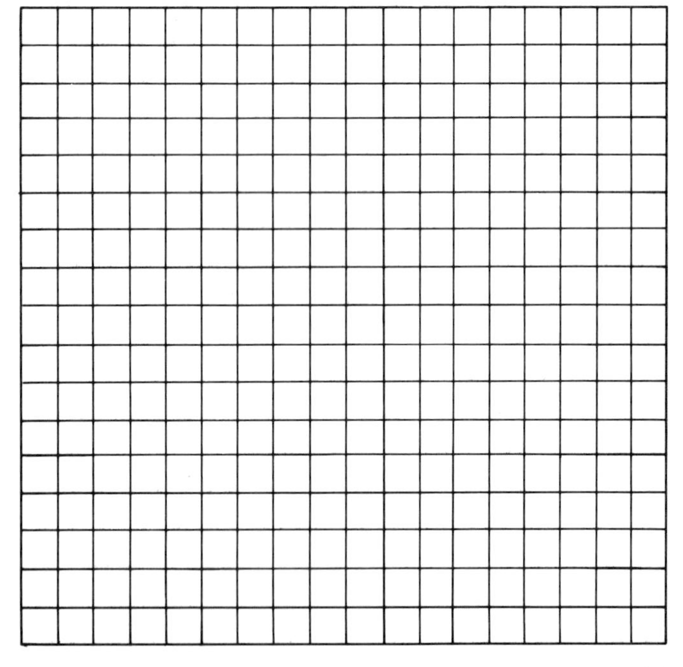

1 Begin by placing tracing paper over the word square grid. Secure it in position with paper clips.
2 Put in as many words as possible. They may be vertical, horizontal, diagonal or even backwards.
3 Write down the words you have used in a list, below the puzzle.
4 Finally, fill in any blank spaces with unusable letters.

Exercise 2

1 Which statement gives the most accurate explanation? Write down the complete correct sentence for each definition.

a) An element is
- A a kind of mineral
- B a substance that cannot be broken down any further chemically
- C made of different compounds
- D found in the Earth's crust

b) An atom is
- A extremely small
- B linked with other atoms in a mineral structure
- C is the smallest part of an element that can exist on its own
- D used to make nuclear bombs

c) The Mohs scale is
- A a way of measuring mineral strength
- B a 10-point scale of mineral resistance
- C a set of scales for measuring mineral mass
- D a set of standard minerals increasing in hardness from 1–10

d) The streak of a mineral is
- A the colour of its powder
- B a way of powdering a mineral
- C method of recording the colour of a mineral
- D the way a mineral reflects light

e) The cleavage of a mineral is
- A the ability of a mineral to split along planes of weakness
- B the way the mineral breaks up
- C an uneven break
- D a fragment that has split from its parent mineral

f) The habit of a mineral is
- A its brilliance and sparkle
- B where it is usually found in the Earth's crust
- C the form or shape
- D the usual size and colour

2 Solve the following mineral riddles.
- a) I am silvery grey, so is my streak; I feel quite heavy. What am I?
- b) I am harder than glass; my streak is white; I have a glassy lustre. I cannot cleave. What am I?
- c) I feel cold when you touch me; I make your hands black and dirty! I am very soft. What am I?
- d) I am as dense as I am hard; I effervesce in acid. I have a pale green streak. What am I?
- e) I have a golden look but do not be fooled! My lustre is metallic; my hardness is 6. What am I?

Exercise 3

1 Write down, in a table, what the minerals diamond and graphite are like. Put the descriptions under two headings: *Diamond* and *Graphite*.

Property	Diamond	Graphite
Composition	Carbon	Carbon
Hardness		
Colour		
Structure		
Possible uses		

2 Read this passage, then answer the questions.
'I went down into the cave. All the walls were wet, shiny and white. Structures like icicles were hanging down from the roof. Water was dripping onto the floor of the cave.'

a) What is the white mineral likely to be?
b) What kind of rock is the cave likely to be in?
c) What is the hardness of the white mineral?
d) Write down two other properties of this mineral that would help you to confirm its identity.

3 You are given pieces of fluorite and halite. Both specimens are white in colour. Explain what tests you would do to tell them apart.

4 a) What is Mohs scale of hardness?
b) Explain the importance of this test in identifying unknown minerals. Give examples in your answer.

5 Explain fully the meaning of the following terms: silicon oxygen tetrahedron vein mineral crystal texture.

6 Explain the criteria that have been used to arrive at the mineral classification in Table 2.

Table 4 The physical properties of the twenty minerals

A = silicates
B = single (native) elements
C = oxides
D = sulphides
E = carbonates
F = halides
G = sulphates

	Group	Hardness	Colour	Streak	Lustre	Density	Cleavage or Fracture	Habit or Form	Other tests
	QUARTZ (silica) SiO_2	7	usually white or transparent	white	glassy	2.7	none conchoidal fr.	crystals or massive	easily scratches glass
	ORTHOCLASE FELDSPAR silicate of K+Al	6	light pink or white	white	glassy	2.6	rectangular	rectangular crystals, or massive	
	OLIVINE silicate of Fe + Mg	6–7	shades of green/brown	colourless	glassy	3.8	conchoidal fr. poor cleavage	massive and granular	
A	AUGITE silicate of Al, Ca, Fe, Mg	5–6	greenish black	nearly white	glassy	3.4	2 sets of cleavage at right angles	crystals or massive	blocky crystals
	HORNBLENDE silicate of Al, Ca, Fe, Mg, Na	5–6	greenish black	nearly white	glassy	3.4	2 sets of cleavage at 120°	crystals or massive	long crystals
	BIOTITE MICA silicate of K, Al, Fe, Mg	2.5	dark brown to black	white	glassy	3	good, flaky thin sheets	thin flakes	flakes are elastic
	MUSCOVITE MICA silicate of K, Al	2.5	transparent in thin sheets, or yellow/brown	white	glassy	3	good, flaky thin sheets	thin flakes	silvery elastic flakes
B	GRAPHITE (C)	1	black	shiny black on paper	metallic	2.1	flaky	greasy flakes or massive	feels cold
C	HAEMATITE Fe_2O_3	5.5	black to red/brown	reddish brown	metallic	5	none. Uneven fr.	reniform	easily oxidises (rust)
	MAGNETITE Fe_3O_4	5.5	iron-black	black	metallic	5.2	poor; subconchoidal fr.	usually granular or massive	magnetic
	GALENA PbS	2.5	lead grey, silvery	lead grey	metallic	7.5	cubic: 3 sets at 90°	cubic crystals or massive	bad egg smell with acid
D	PYRITE FeS_2	6–6.5	golden pale yellow	greenish black	metallic	4.9	none; uneven fr.	cubic crystals or massive fibrous nodules	sparks when struck
	SPHALERITE (zincblende) ZnS	4	brown to black	white/to brown	plastic	4	good rhombic cleavage	small crystals common; also massive	slight smell with acid

(continued)

Table 4

	CALCITE Ca CO_3	3	white or transparent	white	glassy	2.7	rhombic: 3 sets not at 90°	varied crystals or massive	fizzes with acid
E	DOLOMITE Ca, Mg CO_3	4	white, yellow or brown	pale yellow	pearly	2.8	rhombic	rhombic crystals, with curved faces	slight fizz with acid
	MALACHITE Cu CO_3	4	bright green	pale green	silky to dull	4	none uneven fr.	reniform, banded	fizzes with acid
F	FLUORITE (fluorspar) Ca F_2	4	varied: purple, green, yellow, white or transparent	white	glassy	3.2	octahedral	cubic crystals	
	HALITE (rock-salt) NaCl	2.5	white, light brown or transparent	white	glassy	2.2	cubic	cubic crystals	salty taste
G	BARITE (barytes) Ba SO_4	3	white, grey, yellow	white	glassy	4.5	good cleavage in 3 directions	tabular crystals, or massive	feels heavy
	GYPSUM $CaSO_4 + 2H_2O$	2	usually white	white	pearly or silky	2.3	flaky, fibrous	flat crystals; granular or fibrous	soft, crumbly

Table 5

Mineral	Occurrence and main uses
QUARTZ	Found widely in many rock-types. Easily survives weathering and erosion. Quartz grains common in sedimentary sandstones. *Uses:* glass; electrical
ORTHOCLASE FELDSPAR	The feldspars are the most abundant of all minerals – found in many rocks. *Uses:* glazing of pottery, porcelain, ceramics
OLIVINE	The main mineral of many darker igneous rocks. Unstable in weathering – easily breaks down – so rare in sediments. PERIDOT is a gem variety
AUGITE	Short stumpy prismatic crystals. Found in many igneous and metamorphic rocks.
HORNBLENDE	Long blade-like crystals. Found in many igneous and metamorphic rocks.
BIOTITE	'Black' mica. Found commonly as black speckle-like crystals in granites and other rocks. Very abundant in schists and gneisses (metamorphics)
MUSCOVITE	'White' mica. Shiny flakes visible in many rocks. *Uses:* electrical/heat insulator; 'glitter' on cards
GRAPHITE	Dirty to handle; very soft. Feels cold because it conducts heat well. *Uses:* lubricant, paint, mixed with clay for pencil lead
HAEMATITE	An important iron ore, over 70% Fe. Also used as a pigment/polishing powder
MAGNETITE	An important iron ore, 72.4% Fe. Strongly magnetic
GALENA	A common vein mineral found with calcite, quartz and fluorspar. *Use:* The main ore of lead. Silver is an important by-product
PYRITE	Known also as Fool's Gold, because of its colour, although much harder than gold. Can be used to produce sulphuric acid.
SPHALERITE	A vein mineral found with galena and other minerals. The most important ore of zinc. *Uses:* metal alloys; solder; toothpaste tubes
CALCITE	A very common mineral. It may form extensive limestone beds. Also found as deposits in caves. *Uses:* fertiliser; cement; glass
DOLOMITE	Found in extensive beds – usually forms by alteration of limestone (calcite) after deposition. *Uses:* building stone; refractory blocks
MALACHITE	A valuable ore of copper (57% Cu). Also cut and polished for ornamental purposes
FLUORITE	A common vein mineral – with galena, quartz, calcite and others. *Uses:* as a flux to lower the melting point of iron
HALITE	Breaks into tiny cubes. Soluble in water. *Uses:* food preserving; glass and soap-making
BARITE	A common vein mineral containing barium found with galena and other minerals
GYPSUM	Forms in arid conditions of high evaporation. *Uses:* cement-making; Plaster of Paris; plasterboard

8 ROCKS

The great variety of rocks exposed at the Earth's surface can be divided up into three main groups:

1 *Igneous:* rocks with crystals that form by molten magma cooling above or below the Earth's surface.
2 *Sedimentary:* broken fragments of older rocks that collect in layers.
3 *Metamorphic:* rocks that have been altered by great heat and pressure.

Fig 8.1 Basalt, with vesicles. Basalt is erupted from volcanoes. Cooling in the lava is fast and so the crystals formed are small. Vesicles are cavities left by gas.

Fig 8.2 Obsidian – a lava made of glass

1 IGNEOUS ROCKS

These rocks form from molten magma. Magma is a hot solution of chemicals found just below the Earth's crust. The hot liquid rock rises towards the surface, where it is forced out as lava. Lava is an *extrusive* rock. Magma may also cool and harden below the surface as an *intrusive* rock.

When magma cools slowly below ground, the crystals have plenty of time to grow quite large. The time taken for cooling and crystallisation depends on the size of the igneous body. A large granite bathylith could take several thousand years to cool and crystallise. This is why granite is coarse-grained. Dykes or sills are usually much thinner and lose heat more quickly. The crystals in the rock are therefore smaller. One medium-grained rock seen in many dykes and sills is *dolerite*. Surface lavas cool even more quickly. Crystals do not have time to grow large. They may only be visible with a hand lens or microscope, eg, *basalt* (Fig 8.1). In some cases the cooling of lava is so fast that there is no time for crystals to grow at all. When this happens a natural glass is formed, eg, *obsidian* (Fig 8.2).

Columnar jointing

Sometimes, thick lava flows cool quite slowly and forms *columnar jointing* (Fig 8.3). As the lava cools, it

Fig 8.3 Columnar jointing in a thick lava flow: the Giant's Causeway, Antrim, Northern Ireland

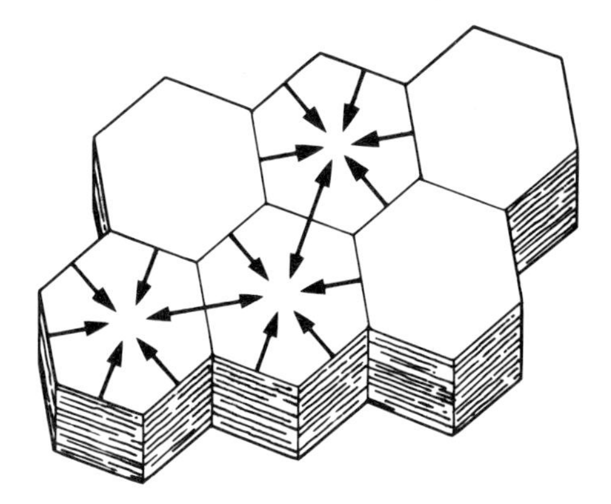

Fig 8.4 Hexagonal columns develop as the lava contracts on cooling, in much the same way as mud cracks when it dries out

contracts. Hexagonal shrinkage cracks develop to form vertical columns from the top to the bottom of the flow. Each column is 20–80 cm across (Fig 8.4).

Crystal texture

Look again at Figs 7.2 and 7.3 (page 42). Can you see how the crystals of different minerals have grown together as the rock has cooled? Why are the feldspar crystals so well formed in contrast to the quartz? Clearly the larger feldspar crystals have had longer to grow in the hot solution. This means that they were the first to crystallise at high temperatures. As the magma cooled, quartz, and finally mica, began to form crystals in the spaces that were left. Therefore they are not so well formed. Early well-formed crystals are called *phenocrysts*. The rock containing them is said to have a *porphyritic* texture (Fig 8.7). A porphyritic basalt will have larger crystals of quartz and feldspar dotted about among much smaller grains.

Activity 1: How igneous rocks form

This activity shows what happens when crystals of different minerals grow in the cooling magma, to form a solid igneous rock.

1 Pour a little water into a beaker and heat it gently in a pan of water (Fig 8.5).
2 Add equal amounts of copper sulphate (a blue mineral) and potash alum (a white mineral). Heat and stir until dissolved. Continue to add the two minerals until no more will dissolve. The solution is then saturated.
3 Put some of the solution onto a cold watch glass to cool quickly. Pour the rest into a beaker and loosely cover. Leave until the following lesson to allow the solutions to cool and crystallise (Fig 8.6).
4 Record your results.
   a) Make a sketch of the crystals in the watch glass and beaker.
   b) Which container has the larger crystals? Why is this?
   c) Which mineral is the first to form crystals?
   d) Which mineral always forms the largest crystals?
5 Compare your results to Figs 7.2 and 7.3 showing the crystal texture of granite.

Conclusions

The minerals which crystallise first in a rock, are well formed. They control the texture of the rock and other minerals have to crystallise in the spaces between them. The slower the rate of cooling the larger the crystals.

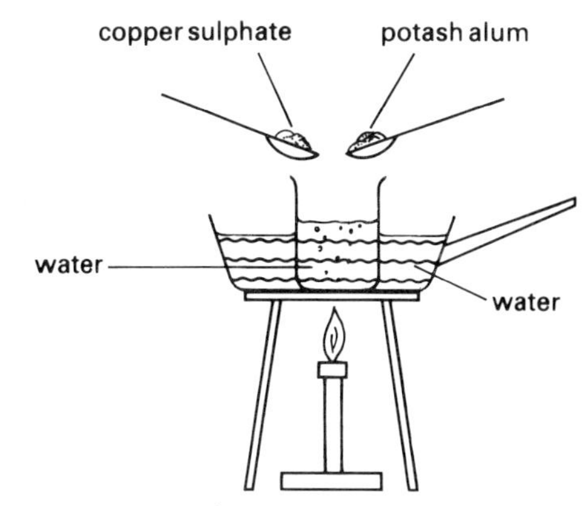

Fig 8.5 Growing crystals

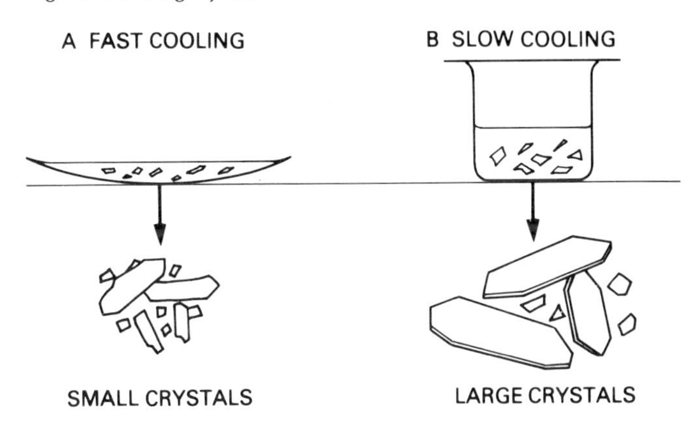

Fig 8.6 The formation of different-size crystals

Fig 8.7 Porphyritic basalt from Arthur's Seat, Edinburgh

Table 1

	Large grains	*Medium grains*	*Fine grains*	*Glass*
Name of rock				
Number of minerals				
Colour of minerals				
Is rock intrusive or extrusive?				
Did the magma cool: *very fast/fast/slowly/very slowly*				

CLASSIFYING IGNEOUS ROCKS

The size of the mineral grains is a good guide to the kind of cooling environment in which the rock formed. It is also a useful way of sorting out rocks. Another way is to look at the silicate minerals present. Quartz, feldspar, augite and hornblende are common in most igneous rocks.

The mineral content affects the colour of the rock. Quartz and the feldspars are light coloured. They are known as *felsic* minerals. Augite and hornblende contain magnesium and iron and are dark coloured. They are known as *mafic* minerals.

Table 2 show the names of some common igneous rocks. Their position in the table depends upon grain size and mineral content. Rocks like granite contain a lot of quartz. They are known as *acid* igneous rocks. Rocks like gabbro have no quartz. They are known as *basic* igneous rocks. Between the two extremes are the *intermediate* igneous rocks, eg, diorite.

Table 2 Igneous rocks

		Mineral content		
Grain size (diameter)	*Cooling environment*	About 30% quartz and feldspar	About 30% hornblende and feldspar	About 50% augite and feldspar
Glassy	very fast, lavas	obsidian (usually dark in colour)		
Fine, less than 1 mm	fast, lavas	rhyolite	andesite	basalt
Medium, 1–5 mm	slow, dykes and sills	microgranite	microdiorite	dolerite
Coarse, over 5 mm	very slow, large intrusions, eg, bathylith	granite	diorite	gabbro
Amount of silica (quartz) present		*Acid*	*Intermediate*	*Basic*
Colour		felsic (light) →		mafic (dark)

Activity 2: Looking at igneous rocks

1 Use your knowledge of grain size and cooling rates to sort out and classify specimens of granite, dolerite, basalt and obsidian.
2 Examine each rock using a hand lens, and make notes on what you can see.
   a) How many different minerals are there in each specimen?
   b) What are their colours?
   c) What shape and size are they?
3 Copy out Table 1 into your book and use it to record your observations about each rock.

Activity 3

1 Look at hand specimens of the different igneous rocks listed in Table 2.
2 Place them on the correct positions on the grid, using grain size and colour as the main guide.

Exercise 1

Write out and complete the following statements using the words listed below.

1 ________ rocks are eroded fragments of older rocks that have been deposited in layers.
2 ________ rocks have formed by the cooling of molten ________.
3 Metamorphic rocks have been ________ by great heat and pressure.
4 Magma is a hot ________ of chemicals.
5 Igneous rock which forms below the surface is called an ________.
6 A mineral has an individual ________ make up and shape.
7 A crystal has smooth flat ________ and a ________ shape.
8 The way in which crystals interlock to form a solid rock is known as crystal ________.
9 Slow cooling produces ________ crystals; whereas fast cooling produces ________ crystals.
10 Rapid chilling of a lava in sea water could cause natural ________ to form. The rock called ________ is one example.
11 Extrusive igneous rocks form at the ________.

texture surface glass sedimentary altered regular magma solution intrusion small, large faces chemical obsidian igneous

Exercise 2

1 a) Look at Fig 8.8. Write down the numbers 1 to 4, and beside each one put the correct description for crystal size: large crystals, medium sized crystals, fine grained, glass.
   b) Explain why the rates of cooling will be different in the four environments.

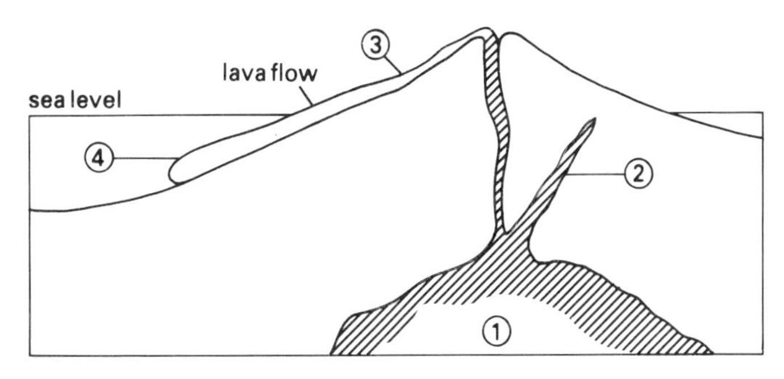

Fig 8.8 Four different cooling environments

   c) What are the names for the igneous structures at 1 and 2?
   d) Name a possible rock type for each of the cooling environments 1–4.
2 A specimen of volcanic rock lava was found: it had quite large well-formed crystals surrounded by many tiny crystals all of the same size (Fig 8.7).
   a) What is the name given to well-formed large crystals?
   b) It was decided that the rock must have had an unusual cooling history. Explain how it must have formed and refer in your answer to Fig 8.8.
3 What is the nature of granite and why are its crystals so clearly visible?
4 a) Describe a hand specimen of:
      (i) basalt with vesicles;
      (ii) obsidian.
   b) Account for the different appearance of these rocks.
7 What is meant by columnar jointing?
   a) Name two places where it might be found.
   b) Explain fully how the structure forms and illustrate your answer.

2 SEDIMENTARY ROCKS

What is sediment?

The action of weather and the forces of erosion produce a constant supply of broken rock fragments. This material is transported by rivers, wind, ice and sea waves. It is then dumped to form layers of sediment.

Most *strata* (layers), form in the horizontal position. Each stratum has a distinct surface, called a *bedding*

Activity 7: Making layered rocks (1)

1 Fill a bottle with aquarium gravel, sand, mud and water. Replace the cap and shake thoroughly.
2 Allow to settle.
3 Note the way the sediment forms layers. What kind of material settles first? Why does the sediment behave in the way it does?

   The water may stay cloudy for some time because fine mud is held in suspension. How long does it take for the fine mud to settle and the water to clear?

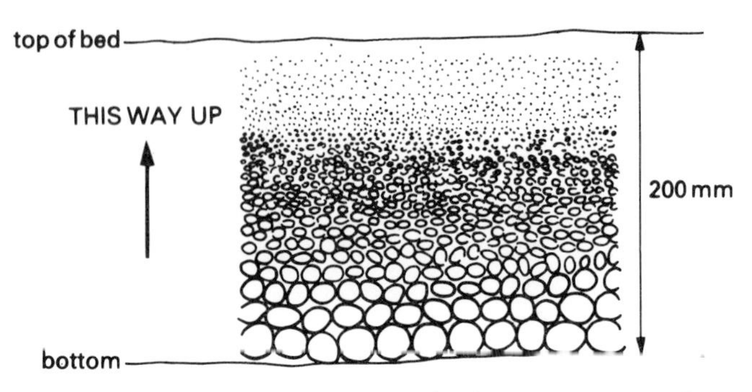

Fig 8.9 Graded bedding – a useful guide to the correct 'way up'. It can tell you whether a set of beds has been overturned.

plane. This separates the bed from other layers above or below it. The bedding plane marks a break in the deposition of the sediment and a change in conditions.

Bedding

Graded bedding
Look closely at beds of sandstone exposed in cliffs. It is often possible to see the coarse pebbles at the bottom of a layer, grading up to fine sand near the top. This is called *graded bedding* and is clear evidence that the rock must have formed in water (Fig 8.9).

Current bedding (cross-bedding)
This kind of bedding is produced by a current of wind or water moving in one direction. The suspended material therefore settles out down a slope (Fig 8.10). Current bedding is very common in sandstones deposited in deltas and in desert sandstones. Can you think of any reasons for this?

Grain size

The size of the fragments is important. This is summarised in Table 3.

When gravel and pebbles are carried by rivers, they often settle out quickly because they are relatively heavy. It is usual to find coarse-grained sediments like this near to the shore or banks, where a rock called *conglomerate* eventually forms (Fig 8.11). The pebbles have often been rounded by movement along the beach or river bed.

Sand settles further out from the shore to form layers of *sandstone* (Fig 8.12). Mud will stay in suspension and may be carried some distance from the land. Eventually it settles to form fine *mudstones* or *shale* (Fig 8.13).

Table 3

Sediment grain size (diameter)	*Name*	*Possible sedimentary rock*
Coarse (over 5 mm)	Boulders, cobbles	Conglomerate
Medium (between 5 and 1 mm)	Pebbles, gravel, sand	Sandstone
Fine (less than 1 mm)	Fine sand, mud, silt	Shale, mudstone

Fig 8.10 Current bedding. The finer current bedding is inclined to the main bedding planes. This structure is also a useful 'way-up' criterion in fieldwork. Which way was the current moving when these rocks were deposited?

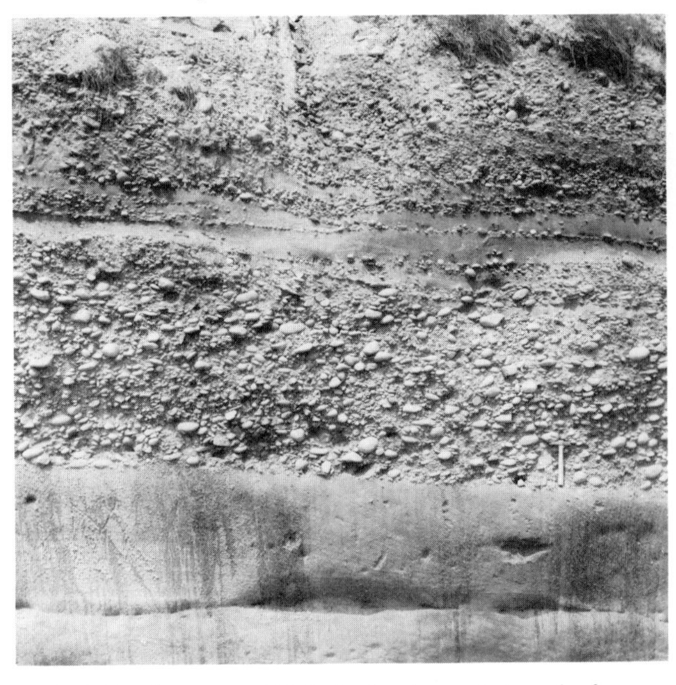

Fig 8.11 Conglomerate pebble beds. Conglomerate is made of pebbles held together with silica, calcium carbonate or clay. The rounded pebbles have been eroded, and were originally part of an igneous or metamorphic rock. (Notice the hammer, showing the scale.)

Fig 8.12 Ripple-marked sandstone. Sandstone is made of layers of sand. Sand is made up of fragments of quartz and other minerals cemented together. Pore-spaces exist between the 'grains'. Shallow water ripple-marks are common on the top surface.

Fig 8.13 Shale – a fine-grained layered mudstone, made of tiny fragments of quartz and feldspar with clay minerals. It is soft and breaks easily. Fossils are often found in shale.

Compaction and cementing of sediments

In time, layers of sediment on lake beds or the sea floor become buried. The weight of newer layers on top of them *compacts* the sediments. Rock grains are pushed closer together until they interlock. Water in between the grains is squeezed out. Minerals that were once dissolved in the water are left as a coating on the grains. This helps to *cement* them together. In this way a sloppy layer of sediment can become a solid rock. Eventually the new rock layers may be folded by earth movements, and could be uplifted to form land again.

Activity 8

Use a land lens to examine rock samples of conglomerate, sandstone and shale.

Note how the fragments are compacted and cemented together. Use a penknife to scratch the rocks to see how hard and solid they are. Can you see any fine fragments falling away from the samples?

Activity 9: Making layered rocks (2)

You will need sand, Polyfilla, colouring agents, Vaseline, shells, three small yoghurt containers (plastic) labelled A, B and C, and one large yoghurt container (plastic).

1. Put equal amounts of damp sand into the three yoghurt containers A, B and C. Add a different food colouring agent to each cup and mix thoroughly. (Fig 8.14)
2. Add one teaspoon of Polyfilla powder to each cup and mix thoroughly. Do *not* add water. The Polyfilla will act as a cement to bind the sand together.
3. Transfer mixture A to the large container. Press down the mixture to *compact* the sand.
4. Smear a small shell with vaseline, then press it into the surface of mixture A. The shell represents a fossil.
5. Empty mixture B on top of A and press down. Then add mixture C and do the same. Shells can also be added at these stages.
6. Allow to set and harden.
7. Peel away the plastic container and examine the 'rock' layers. Make a labelled sketch to show: rock-layers, bedding planes, packing of sand grains and the fossils. Compare your 'rock' with actual sandstone. Does it have the same texture?

NB This method could be used to make a range of rock types. You could easily make a conglomerate, by mixing gravel, sand and Polyfilla together, with a little water.

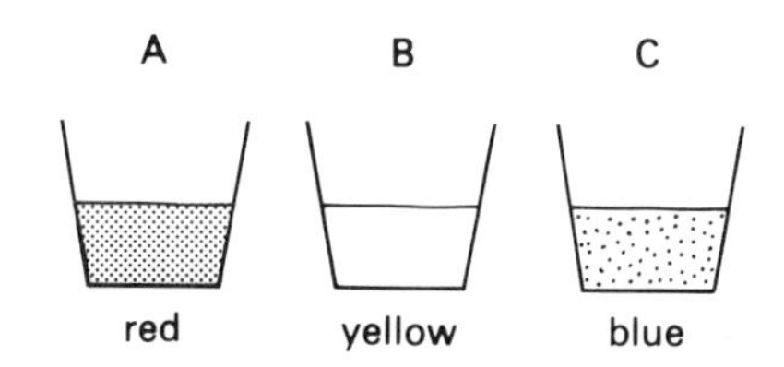

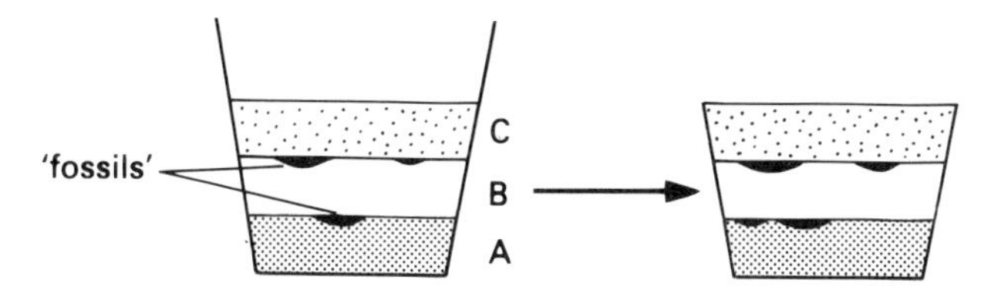

Fig 8.14 Making layered rocks

E

Fig 8.15 Well-bedded Magnesian limestone in a sea stack

Fossils

Many sedimentary rocks contain *fossils*. When plants and animals die, their shells and other hard parts may remain and get trapped in the layers. Shells, bones and corals often become preserved in this way. They provide important evidence of life in the past.

Organic and chemical sediments

Although fragments of previously eroded rocks form the bulk of sedimentary rocks, there are two other distinct types. These are the *organic* and *chemical* sediments.

Organic

Some sedimentary rocks, like coal and many limestones, form because of the activity of plants and animals. They are *organic* in origin. Fragments of shells, corals and other material, often collect in thick layers, to form *shelly limestones*. Many sea animals can extract calcium carbonate from seawater. They build their shells and other skeletal parts from it. When they die their skeletons form layers of lime-rich mud. This has a high calcium carbonate (calcite) content. The mud eventually compacts to form solid limestone rock. Huge structures like the Great Australian Barrier Reef could some day become limestone.

Chemical

In certain situations it is possible for calcium carbonate to be directly deposited from seawater, eg, Magnesian limestone (Fig 8.15). Normally calcium carbonate stays dissolved but in warm seawater it can come out of solution as a solid mineral. Exactly the same thing happens when 'hard' water is boiled in a kettle. Calcium carbonate is deposited as 'fur' on the inside of the kettle.

Exercise 1

Write out and complete the following statements using the words listed below.

1 Rivers, wind, ice, and the sea are all responsible for ________ material.
2 Fine mud remains ________ and is the last to settle out.
3 Layers of sediment become ________ by the weight of new layers above them.
4 Different mineral substances can bind the rock ________ together.
5 A ________ is a rock containing pebbles cemented together.
6 Shale is a finely ________ rock.
7 ________ bedding is the result of coarse material settling before the finer material.
8 When plants and animals die, their remains can form ________.
9 ________ and ________ are both examples of organic sediments.

coal transporting conglomerate compacted limestone grains layered suspended graded fossils

---

Exercise 2

1 Describe fully the process of formation of a solid rock from 'sloppy' sediment.
2 Select from the list below the rock which fits *each* of the descriptions given:
shale conglomerate sandstone shelly limestone.
   a) A rock of pebbles cemented together.
   b) A rock composed mainly of quartz grains.
   c) A rock light grey in colour, containing fossils.
   d) A dark, finely-layered soft rock.
3 Explain why clay tends to settle in quiet, still water and not in rougher conditions such as at sea near the shore.
4 Explain with the aid of *one* labelled diagram how conglomerate, sandstone and clay can be formed near to one another at the same time.
5 a) Describe, with the aid of a diagram, what is meant by 'graded bedding'.
   b) On a field trip, you see the following graded beds of sandstone in a quarry (Fig 8.16).

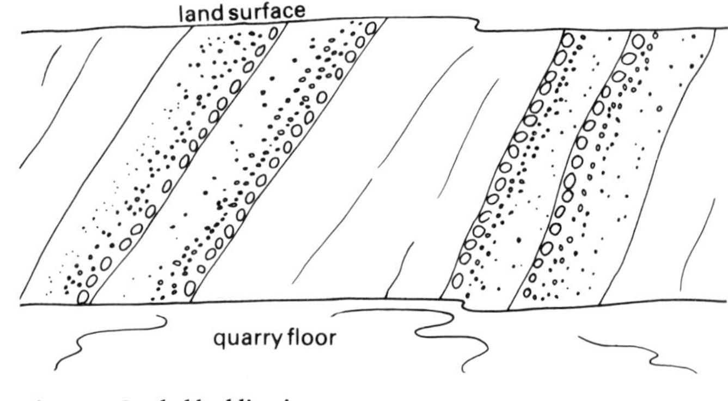

Fig 8.16 Graded bedding in a quarry

(i) Copy the diagram and label the top and bottom of each bed.
(ii) Which way up are they?
(iii) How does the structure of graded bedding help you to say what has happened to these beds?
c) Make a sketch based on Fig 8.16 to show what current bedding would look like in the same situation.

3 METAMORPHIC ROCKS

Metamorphism is the process by which any existing rock is changed by *heat* and *pressure*.

New minerals and crystals are formed within the solid rock. It does not become a molten liquid but rather it undergoes a kind of 'baking' process. Metamorphism is *never* a surface process. It happens deep within the Earth's crust. If the altered rock was a sediment then often the original layers are lost. Any fossils are destroyed or badly distorted, as the rock is heated, and squeezed or stretched (Fig 8.17).

What do metamorphic rocks look like?

Like igneous rocks, metamorphic rocks are made of interlocking crystals. This makes the rocks hard and strong. Also many metamorphic rocks have their minerals rearranged in parallel sheets or bands. This gives the rock a layered appearance (Fig 8.18).

Contact metamorphism by heat

One situation in which metamorphic changes occur is where a magma is injected into sedimentary rocks. The intrusion loses heat to the surrounding rocks as the magma cools. The greatest alteration will be in rocks next to the intrusion. The width of the *aureole* (zone of change) can vary but is rarely more than 300 metres (Fig 8.19). Limestone is easily changed to *marble*. Sandstone does not react so easily to heat. There is only a slight change in texture. The quartz grains fuse together, to form a much harder rock called *quartzite*. The spaces between the sand grains disappear. Shale and other clay rocks also alter. New crystals form and the rocks become harder, forming a new rock called *hornfels*.

Some of the main changes are shown in Activity 10.

Fig 8.17 The destruction of fossils,flattened

Fig 8.18 Banded garnet gneiss

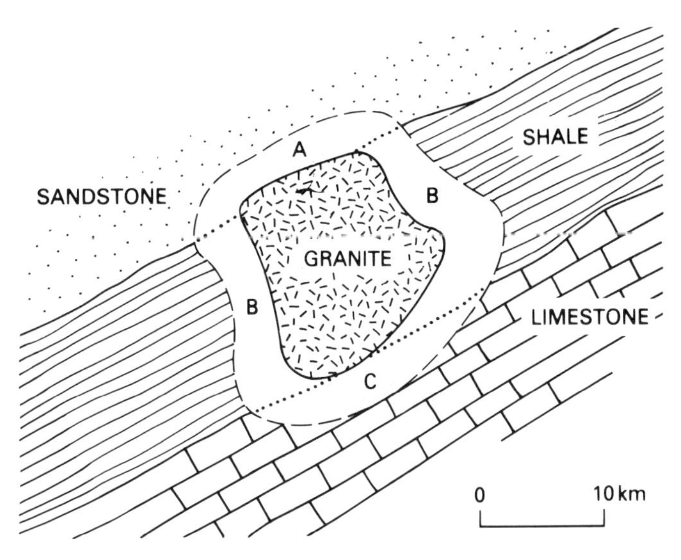

Fig 8.19 Metamorphism around a granite intrusion. The zone where alteration takes place is called the 'metamorphic aureole'. (A, B and C represent new metamorphic rocks – see Exercise 3 Question 3)

Pressure metamorphism: shale to slate

Slates are formed under high pressure conditions. The flaky clay minerals that make up a soft shale or mudstone are squeezed by intense earth movements.

They rotate and line up in one direction to produce thin sheets of rock.

Slate is much harder than shale but easily splits into thin sheets along the *cleavage planes*. This is why slate has been used so much in the past as roofing material.

Regional metamorphism

This kind of metamorphism affects huge volumes of rock at a time. The changes are caused by great heat and pressure. This happens very deep in the Earth's crust, below uplifting fold mountains near to plate boundaries. Here the pressure and intense heat cause many changes. New minerals form, especially in clay rocks which are converted to banded and often intensely-folded rocks known as *gneiss* and *schist* (Figs 8.18 and 8.21).

Temperatures at depth may be great enough for the rocks to melt. If this happens, new magma will form, which then rises to form intrusions and surface lavas.

Activity 10: To show how plastic clay is changed by heat

1 Mould a piece of clay into a brick shape approximately 6 cm × 3 cm × 2 cm.
2 Stand the brick in a dish of water. Does it soak up any of the water?
3 Remove the brick from the dish and allow it to dry out slowly, eg, above a radiator. Then measure its length. Is it smaller? If so, can you think of a reason?
4 Fire (heat) the brick in a kiln for at least three hours to red-heat.
5 Allow it to cool. What colour is the brick? (This colour change is caused by oxidisation of iron in the clay).
6 Measure the brick again. Is there any more shrinkage?
7 Stand the brick in water again. Does it soak up any of the water?
8 Crush the brick to powder. Then add a little water to the powder. Is it possible to make a plastic clay again?
9 Copy this table into your book and use it to record the main changes you have observed.

Activity 11

To show how minerals in a solid rock like slate, can move and respond to pressure.

1 Roll out a layer of plasticine about ½ cm thick. Scatter silver glitter all over and lightly press into the surface.
2 Cut out several equal-sized pieces of plasticine and place them one on top of the other.
3 Square off the layers with a sharp knife. Can you see flakes of glitter?
4 Using two pieces of wood, lightly greased, squeeze the layers (Fig 8.20).

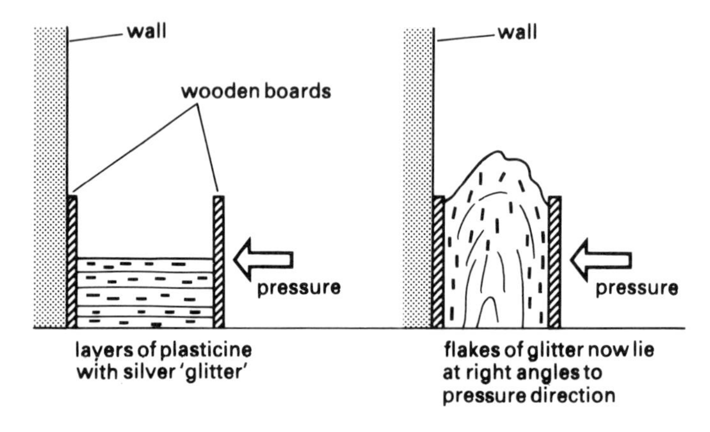

Fig 8.20 Experiment to show how minerals, within rock, respond to pressure

5 Remove the boards and examine the surfaces. Can you see how the flaky pieces of glitter now mostly lie parallel to the boards? Also notice how the original layering has begun to disappear.

Conclusion

What does this activity tell you about alteration in solid rock layers? Write notes about the changes you have seen.

Changes	*Clay brick*	*Fired brick*
Size		
Hardness		
Colour		
Water absorption		
Plasticity		

Fig 8.21 Crumpled schist

Activity 12

The aim of this activity is to record the changes that have been caused by metamorphism.

1 Using a hand lens, examine samples of the following pairs of rock:
   a) sandstone and quartzite
   b) limestone and marble
   c) shale and slate.
2 Record the differences you observe on a copy of the chart below.
3 Summarise the main changes that you observe in each case.

Activity 13: Looking at schist and gneiss

1 Using a hand lens, study a sample of mica-schist. This rock was once a shale. All the clay minerals have been altered to shiny flakes of mica. This gives the whole rock a 'glittering' look.
   a) Look carefully at the flakes of mica.
   b) Are they lined up in one particular direction?
   c) What does this tell you about the pressure direction?
2 Make a sketch of the rock sample.
3 Study a sample of banded gneiss. You will notice that minerals of different composition have separated into light and dark coloured bands. These bands have nothing to do with the original bedding.
   a) Make a sketch of your rock sample. Label the light and dark bands.
   b) Can you recognise any minerals? (Quartz and feldspar are common in the lighter bands.)

THE ROCK CYCLE

Fig 8.22 summarises the slow changes that take place from one rock type to another. Magma rises to form igneous rocks. The exposed surface rocks are worn away by weathering and erosion. The rock grains are deposited in layers, to form sedimentary rocks. They may be folded, uplifted, or metamorphosed. If they melt completely, then new magma forms. The complete cycle takes millions of years.

Rock breakdown: weathering and erosion

The rocks exposed at the Earth's surface are slowly being broken down and dissolved. This is due to the weather, and the effects of air and water. The process is called *weathering*. Weathered rocks are usually much

Grain size (coarse, medium, fine)	Sandstone	Quartzite	Limestone	Marble	Shale	Slate
Texture (make a sketch)						
Hardness (scratch each one with a penknife blade). Is it soft/hard/very hard?						
Colour						

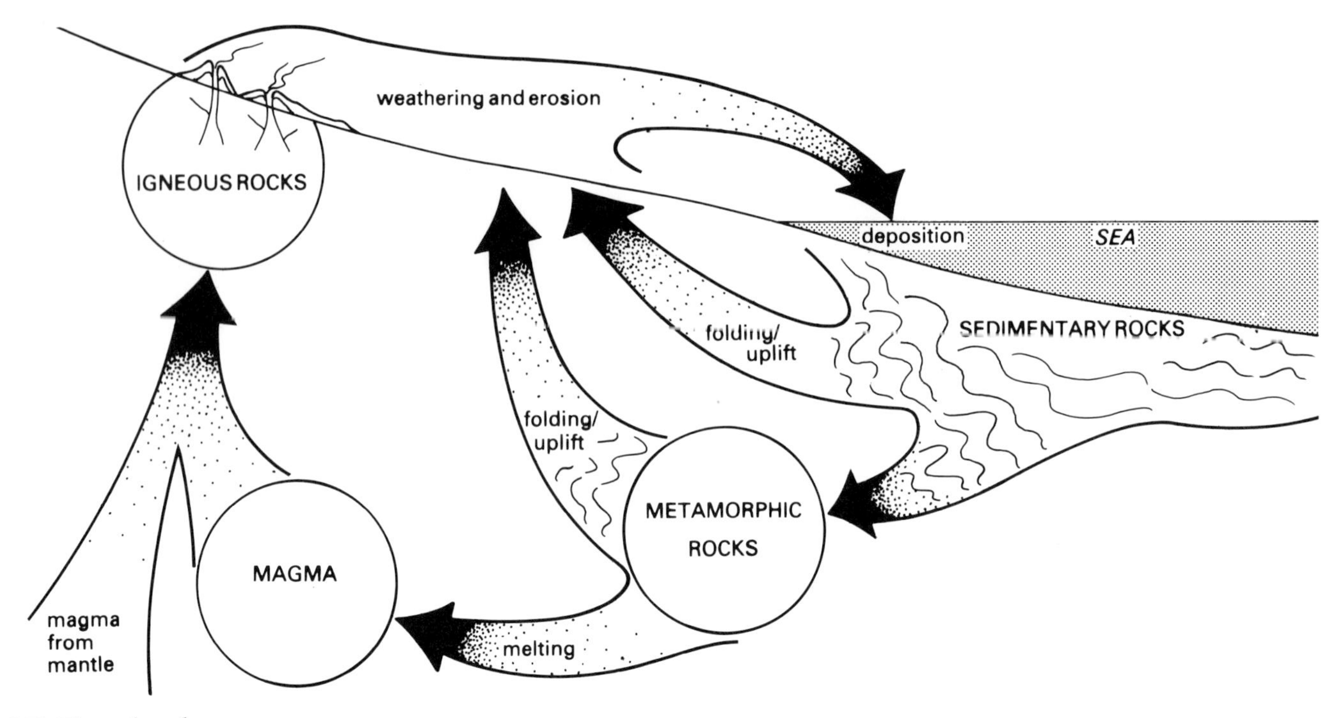

Fig 8.22 The rock cycle

softer than 'fresh' unweathered rocks because the original mineral structures have been destroyed. Therefore they are worn down much more easily by moving water, ice, wind, and wave action. The wearing down process is called *erosion*. Detached rock fragments and grains of minerals, are often carried to the sea. Here the fragments collect in layers. In time these deposits compact to form new sedimentary rocks.

How weathering affects rocks

There are two main processes at work in weathering.
Disintegration This is the breaking apart of rocks by physical forces.
Decomposition This is the decay of rocks by the dissolving chemical action of water on rock minerals.

It is easier to understand this process if you think of how to eat a mint. You can either crunch it with your teeth (disintegration), or dissolve it away by sucking (decomposition). In practice you will probably use both methods on the same mint. The same is true in weathering. Both processes act together as the rock is broken down.

Weathering by disintegration

Frost shattering
In upland areas in cool climates, rocks disintegrate easily when exposed repeatedly to frost action. Water enters cracks and crevices in the rock surface. Ice forms when the temperature drops below freezing. Ice has a larger volume than water and it exerts tremendous pressure on the rocks on either side of the crack and thus widens the cavity (Fig 8.23). During the day the ice may melt. More water collects in the crack and the process repeats itself. The rock eventually becomes fragmented and shattered. This process is known as *frost shattering* or 'ice-wedge' action. Loosened rocks fall down slopes to form piles of loose rocks and pebbles called *scree* (Fig 8.24).

Activity 14: To show how ice expands and breaks up rock

1 Completely fill a plastic lemonade bottle with water.
2 Place upright in freezer and leave to freeze solid.
3 Remove the bottle carefully and observe how it has fractured.

Conclusion
The bottle shatters as the ice expands. Rocks are broken up in a similar way by frost action.

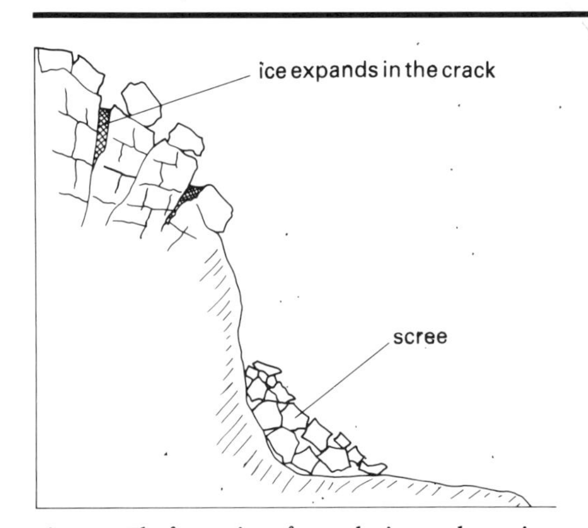

Fig 8.23 The formation of scree by ice-wedge action

Fig 8.24 The screes of Wastwater in the Lake District, England

Heating and cooling

In extreme climates direct heating and cooling of an exposed rock helps to break it apart. The surface of the rock expands when heated by the sun. Rock below the surface remains cool and does not expand. At night the temperature drops sharply and the surface layer contracts. Dew condenses at night and soaks into the surface. The expansion and contraction of minerals at different rates sets up immense stresses. This is coupled with the solvent action of water. The surface layers crumble and disintegrate (Fig 8.25).

Weathering by decomposition

Water is very good at dissolving most of the minerals in rocks. In other words, it is a very effective *solvent*.

Activity 15: To show how different minerals expand and contract at different rates

You will need a bunsen burner and a bi-metal strip (the strip consists of sheets of iron and brass welded together. The metals represent two different minerals).

1 Hold the bi-metal strip with tongs and heat it in the flame of the burner. Record what happens.
2 Remove the strip from the flame. Record what happens. Think carefully about different rates of expansion in different materials, then answer these questions.
   a) Why did the bi-metal strip bend when heated?
   b) Which metal expanded the most?
   c) Why did the strip straighten when removed from the heat?

Conclusion

Different minerals in a rock will also expand and contract at different rates. This process will break up the rock.

When a sugar lump is dropped into a cup of tea, it dissolves because of the solvent action of the hot tea. The sugar goes into *solution*. You can tell the sugar is still there, of course, because of the sweet taste. Rainwater picks up carbon dioxide from the air and from the soil, and it is turned into a weak *carbonic acid*. This speeds up the chemical action of water on rocks, for example, rainwater alters exposed limestone and dissolves it away. (Also see Chapter 13.)

Some of the minerals in rocks dissolve quickly, others take much longer. Consider the three main minerals found in granite. Feldspar is unstable, it changes the most quickly, to a very soft white clay, followed by mica. Quartz is the most stable and takes much longer to change. The soft clay in weathered granite is easily eroded, and the quartz is washed out to become sand grains.

Activity 16: To show how limestone decomposes

1 Prepare two boiling tubes, labelled A and B, and place a small piece of limestone in each tube.
2 Half-fill tube A with water, and tube B with dilute (0.5 M) hydrochloric acid.
3 Watch each tube and see what happens.
   a) Which piece of limestone starts to dissolve?
   b) What kind of reaction is there?

Conclusion

Limestone readily dissolves in water, especially if the water is slightly acid. This process can cause the formation of underground caves as limestone decomposes and is carried away in solution. This process is explained more fully in Chapter 13.

Fig 8.25 Half Dome, Yosemite, California. The surface layers of weathered granite crumble and break away, exposing 'fresh' rock to the same process

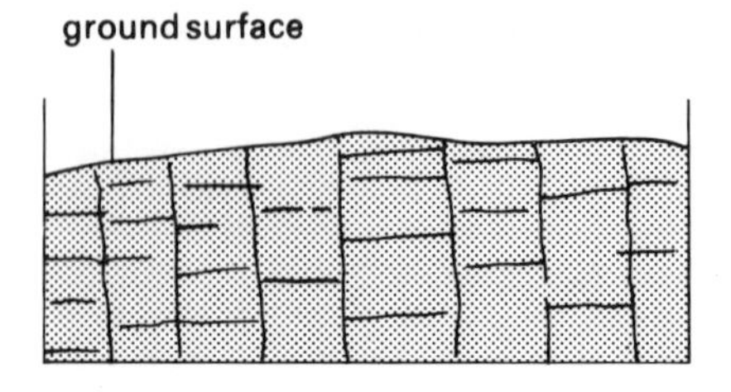

Joints form in the granite as the rock cools and contracts. Then erosion removes overlying rocks.

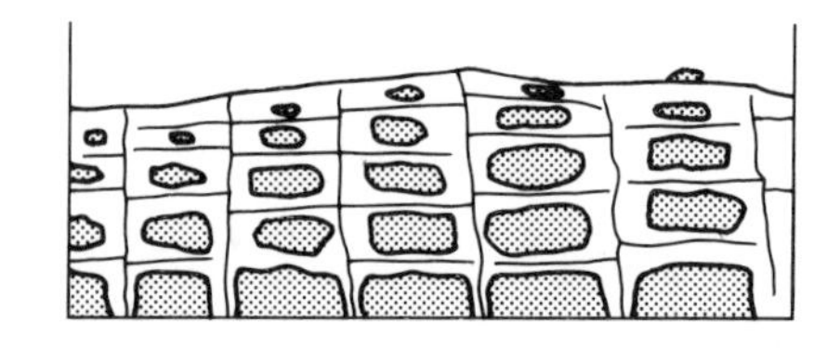

The granite decomposes as water attacks the rock along the joint planes.

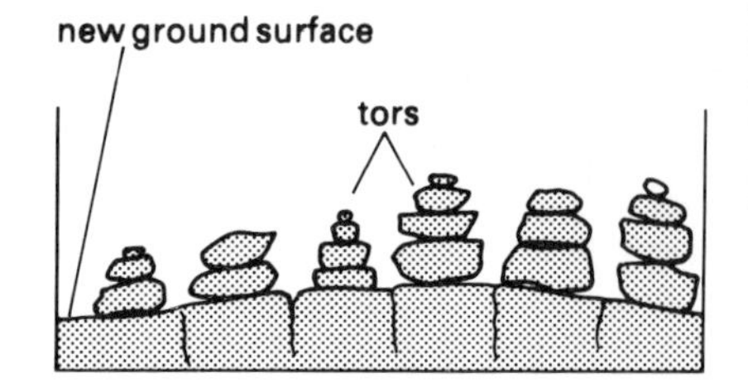

The softer weathered granite is eroded away to leave the stacked blocks of granite as 'tors'.

Fig 8.26 The formation of granite tors

Formation of granite tors

Exposed on the granite moors of south-west England are pinnacles of bare rock known as *tors*. These have formed through the selective weathering of the joints that run through the granite. Water works its way into the joints and slowly attacks the rock. Feldspars are converted to clay, and the rock becomes softened. Eventually the softer weathered rock is washed away. The harder 'cores' of granite are left standing above the lowered ground level (Fig 8.26).

The chemical reactions involved in the breakdown of rocks are often very complex. For example, the reaction of ground water with feldspar in granite may be summarised like this:

$$6H_2O + CO_2 + 2KAlS_3O_8 = Al_2Si_2O_5\,(OH)_4 + 4SiO(OH)_2 + K_2CO_3$$

rainwater (carbonic acid) + orthoclase feldspar = clay mineral + silicic acid + potassium carbonate (in solution)

Exercise 1

Complete the following statements using the words below.

1 The breakdown of rocks in air and water is called ________.
2 The removal and transport of broken up rocks is called ________.
3 Rock fragments collect in ________ on the sea floor.
4 Disintegration is the breaking apart of rocks by ________ forces.
5 Decomposition is the ________ of rocks by chemical action.
6 The breakdown of rocks by ice action is known as ________ shattering.
7 When water freezes and ________ the crack in the rocks is widened.
8 Broken off rock fragments ________ to the foot of the slope, and are known as ________.
9 If a rock surface is alternately heated and cooled, the expansion and ________ causes stresses within it.
10 Water is a very good ________.
11 ________ is the most chemically stable mineral in granite. It does not ________ easily.
12 Rainwater picks up carbon dioxide and forms a weak ________ acid. This is very effective in ________ many rocks, especially limestones.

scree quartz fall weathering decay carbonic erosion dissolving physical frost decomposing solvent layers contraction expands

Exercise 2

1 What is meant by 'frost shattering'? Explain fully how this process breaks down rocks.
2 Describe an activity which shows what happens in the disintegration of rocks by expansion and contraction. Illustrate your answer with sketches.
3 Explain why sand and clay is the usual result of the weathering of a granite mass.
4 Explain the effects of rainwater (weak carbonic acid) on the Earth's surface.
5 Starting with the intrusion of granite into the rocks of south-west England explain the formation of granite tors. Illustrate your answer.

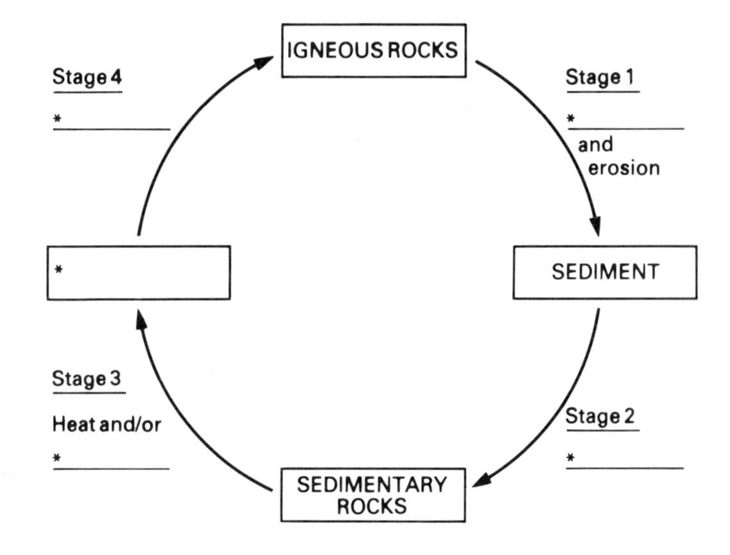

Fig 8.27 The rock cycle

SUMMARY EXERCISES

Exercise 1: On the rocks

(*Solution on page 143*)

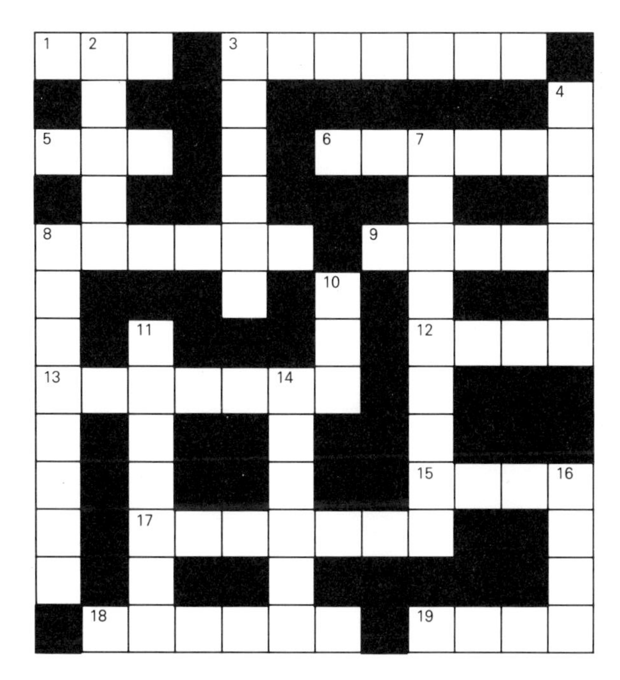

Across

1 Most sediments settle in this (3)
3 An igneous rock commonly found in bathyliths (7)
5 It is like this deep in the Earth's crust (3)
6 Metamorphism will ________ most kinds of limestone (6)
8 This binds the sand grains together (6)
9 Phenocrysts are ________, well-formed crystals (5)
12 Some highly metamorphosed rocks will do this and become magma (4)
13 The name given to a zone of alteration surrounding an igneous intrusion (7)
15 Pinnacles of rock seen on Dartmoor (4)
17 A rock that forms from a hot liquid (7)
18 Banded but metamorphic (6)
19 A very large intrusion will ________ slowly (4)

Down

2 Rivers ________ rocks and transport the fragments (5)
3 Wheat and sand have this in common (6)
4 Hot magma will ________ itself along fissures in the surrounding rocks (6)
7 Another term for small pieces of broken rock (9)
8 It is possible to split slate along these planes (8)
10 When this expands it can shatter rocks (3)
11 The term given to the wearing away of rocks (7)
14 Most sediments are deposited in ________ (6)
16 The name given to an igneous intrusion (4)

Exercise 2

Write out and complete the following statements using the words listed below.

1 A metamorphic rock is formed by alteration due to ________ and ________.
2 Metamorphic rocks only form ________ below ground.
3 Fossils and the original layering are often ________.
4 The alteration zone around an igneous intrusion is called the metamorphic ________.
5 Cleavage ________ form in slate, as the flaky minerals line up in one ________.
6 Sandstone is altered to ________.
7 ________ is altered to slate.
8 In ________ metamorphism large areas of rock can undergo changes.
9 In schist all the ________ minerals have been altered to shiny flakes of ________.
10 In ________ gneiss the lighter coloured minerals have separated from the darker ones.

heat clay regional deep pressure quartzite shale banded aureole mica planes destroyed direction

Exercise 3

1 Describe briefly the two main ways in which metamorphic rocks can be formed.
2 Define 'contact', 'pressure', and 'regional' metamorphism. Select an example of a rock formed in each way from the list below:
gneiss granite shale oolite slate marble.
3 a) Copy Fig 8.19 and make a key to show the metamorphic rocks you would expect to find at A, B and C.
b) Briefly explain how each of these metamorphic rocks has changed from its parent rock.
4 a) What is meant by 'slaty cleavage', and how does it develop?
b) What are the main differences between shale and slate?
5 Describe briefly hand specimens of:
a) schist
b) gneiss.
Explain how they form. Make sketches to show the main mineral textures visible.

Exercise 4

1 Fig 8.27 shows the rock cycle.
a) Copy out and complete the diagram by inserting labels in the five spaces provided.
b) Explain fully what happens to rocks in Stage 1.
c) What are the main processes that occur in Stage 2?
d) Follow the passage of a grain of quartz from an igneous rock, like granite, through the cycle. Name different rocks that would contain quartz at each stage.

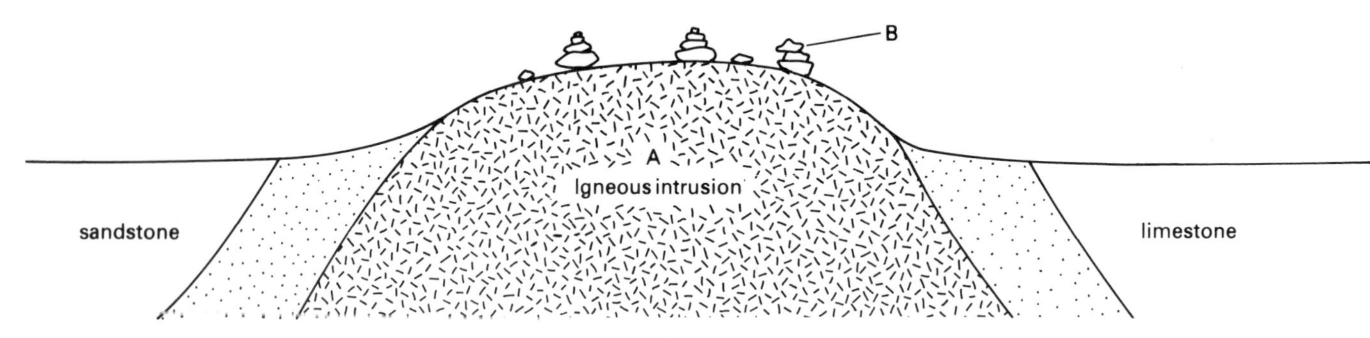

Fig 8.28 A section across Dartmoor, south-west England

2 Fig 8.28 shows a section across Dartmoor in south-west England.
   a) Name the kind of rock found at A. What sort of igneous intrusion is this?
   b) Explain the formation of an upland rocky area at B.
   c) Name two sedimentary rocks that result from the weathering of rock A.
   d) Name the shaded zone on either side of the intrusion and the kind of rocks that would form in this zone.

3 The following account was written by a pupil after a field trip to a quarry. It contains seven *italicised* mistakes. Write out the passage and correct the pupil's mistakes.

I walked along a coastal cliff made of horizontal sedimentary layers. Some of the layers had vertical cracks running through them. I labelled these on my sketch as *bedding planes*. Water had worked its way into these cracks and had *contracted* as it cooled to form ice. This process had obviously weakened the cliff. Loose rocks had fallen off to form *schist* at the foot of the cliff. In one part there was an igneous *extrusion*, one metre across, that cut through the layers at a sharp angle. This I labelled as a *sill*. It was made of a dark, fine-grained rock called *hornfels*. I found a rock called conglomerate containing large rounded fragments. I labelled these as *phenocrysts*.

9 SOIL

SOIL FORMATION

Soil formation is the slow development of a fragmented surface layer. This is derived by weathering from the rocks beneath. It takes about 200 years to form a layer of soil one centimetre in depth. Soil is the vital link between solid rock and living things. Most plants cannot survive without it.

Weathering

The exposed surface rocks are broken up by weathering processes. As a rock is fragmented, more and more of the surface area is affected by the solvent chemical action of water (Fig 9.1). Minerals in the rock dissolve and new clay minerals are formed. In time a hard rock becomes softened and decomposes more easily to soil.

Plants

Plants play an important part in the soil-forming process. The roots of plants force their way into cracks and along bedding planes. This helps to break up the rock. Plants called *lichens* can grow on bare rock. They do not need soil. Lichens produce acids which help to dissolve the rock and mineral salts are released that can be absorbed by the lichens. Other simple plants with roots, such as *mosses*, live on some rocks and benefit from the minerals released by the work of lichens.

When plants die, their decayed remains add important organic substances to the soil. Decaying organic matter is called *humus* and gives soil its dark brown colour. Humus is broken down by the activity of *moulds* and *bacteria*. These are *micro-organisms*, very small animals and plants. The breakdown of humus releases many *nutrients* that plants need, such as nitrogen.

Plants speed up the formation of soil but a good plant covering also protects soil from erosion. The roots bind the soil together so wind and water remove the soil more slowly. The thickness of a layer of soil is the difference between the amount produced by weathering and the amount removed by erosion.

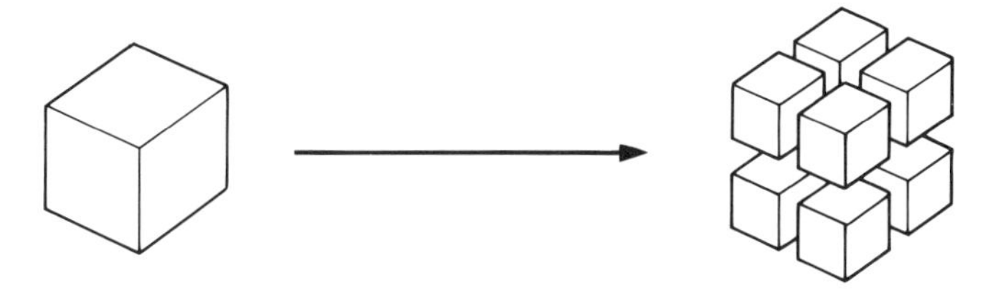

Fig 9.1 Fragmented rocks have a greater surface area exposed to the solvent action of water. If a cube (sides 1 cm) is cut into eight smaller cubes, what will be the increase in surface area?

Animals

Many animals are attracted to rotting plant material, eg, slugs, earthworms, millipedes and woodlice. Other animals like centipedes, spiders and mites, feed on the plant-eaters. Animal movement and activity greatly improves soil structure and assists the breakdown process. Earthworms and other burrowing animals help to improve soil fertility. Their burrows allow air and water, essential for healthy root growth, to enter the soil. Earthworms also eat soil. As it passes through the gut of the worm, the soil is broken down into finer particles; more useful mineral salts are released.

Careful measurements have shown that an average hectare (10 000 m^2) of soil may contain 375 000 earthworms. The worms rework the soil and thirty tonnes of very fine soil is brought to the surface as worm casts each year.

THE SOIL PROFILE

As soils slowly form, distinct layers called *horizons* develop. A cross-section of a soil is called a *soil profile* (Fig 9.2). The thickness of the profile depends on the rate of formation and erosion. Some soils are very thin, others may be up to 10 metres deep. A soil with well-developed horizons is called a *mature* soil. Poorly developed soils are called *immature* or *skeletal*.

The 'A' horizon is the *topsoil* layer. It contains most of the plant and animal life. It is likely to be darker and richer-looking than the lower layers of soil because of its humus content. Horizon 'B' is the *subsoil*. It contains layered rock fragments, often mixed with finer clay washed down from above. It is much more compact than topsoil, with fewer air spaces. It is more likely to be waterlogged and contains little plant and animal life. Horizon 'C' is the bedrock layer. It is slowly weathered to produce more subsoil.

DIFFERENT SOILS

The type of soil which forms depends on the kind of rock fragments it contains. This in turn depends on the type of weathered parent rock. Some examples are shown below:

sandstone → sandy soil
granite → sandy soil (quartz fragments)
chalk → lime soil
shale → clay soil.

Texture

The texture of soil depends on the size of the weathered rock grains. These may vary from coarse gravel to finest clay.

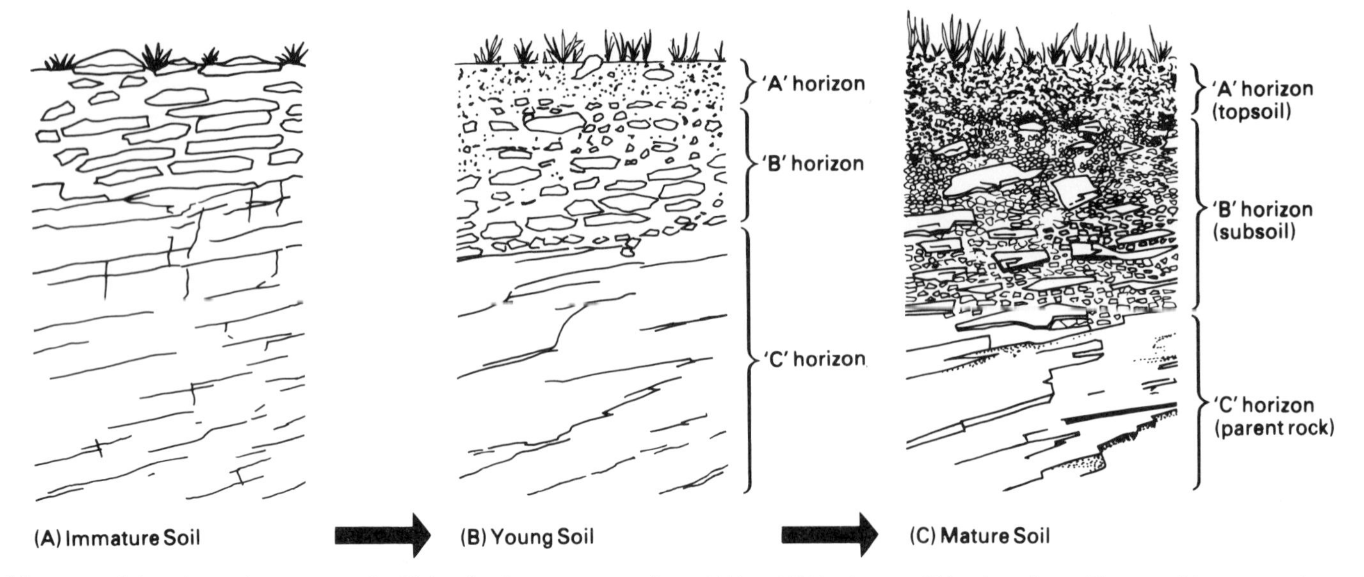

After several decades an immature soil will develop into a young soil, and 'A' and 'B' horizons will begin to form. The transition to a mature soi is likely to take hundreds of years of weathering.

Fig 9.2 Development of a soil profile

Grains	*Size* (diameter in mm)
Gravel	larger than 2 mm
Sand	2 mm – 0.02 mm
Silt	0.02 mm – 0.002 mm
Clay	smaller than 0.002 mm

Texture affects the following properties:
a) the amount of water in the soil;
b) the rate at which water drains through soil;
c) the amount of air in the soil.
Soil grains are covered with a film of water and there are air spaces between the grains (Fig 9.3). There are more air spaces between the grains in coarsely-textured soils. In finely-textured soils, the small grains are more closely packed and there are fewer air spaces.

Clay and sandy soils

Finely-textured clay soils are rich in plant foods but drainage is poor and they easily become waterlogged. The water makes the soil heavy and difficult to cultivate and cuts off the air supply to the roots of plants. Heavy clay soils can be improved by adding sand and humus. This improves the texture and drainage. Lime can also be added to make fine clay clump together into larger particles. This is called *flocculation*.

Coarsely-textured sandy soils are lighter and easier to cultivate with plenty of air spaces. However they dry out very quickly because water can drain through them rapidly, and plants may suffer in dry weather. Plant foods also tend to be lost. They are carried down by the draining water. This process is called *leaching* and can make a soil *infertile*.

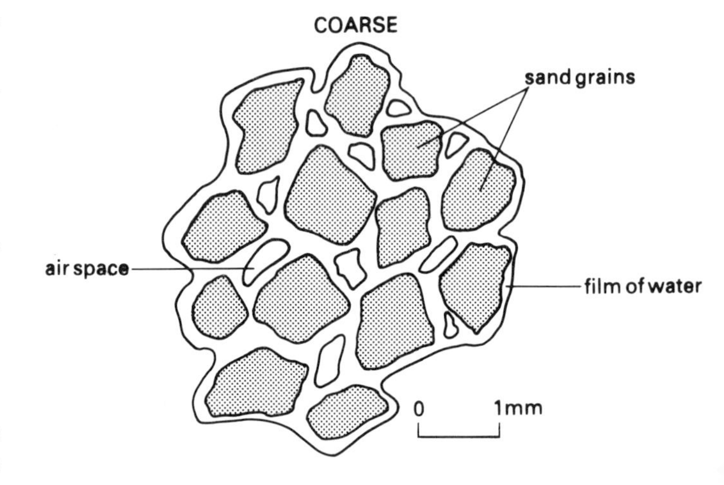

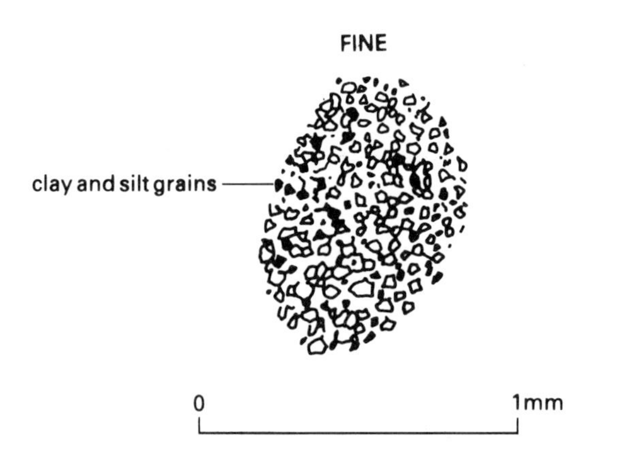

Fig 9.3 Soil texture

The best kind of soil for cultivation and plant growth is a mixture of sand and clay, called a *loam*. This combines all the advantages of coarse and fine soils without their disadvantages.

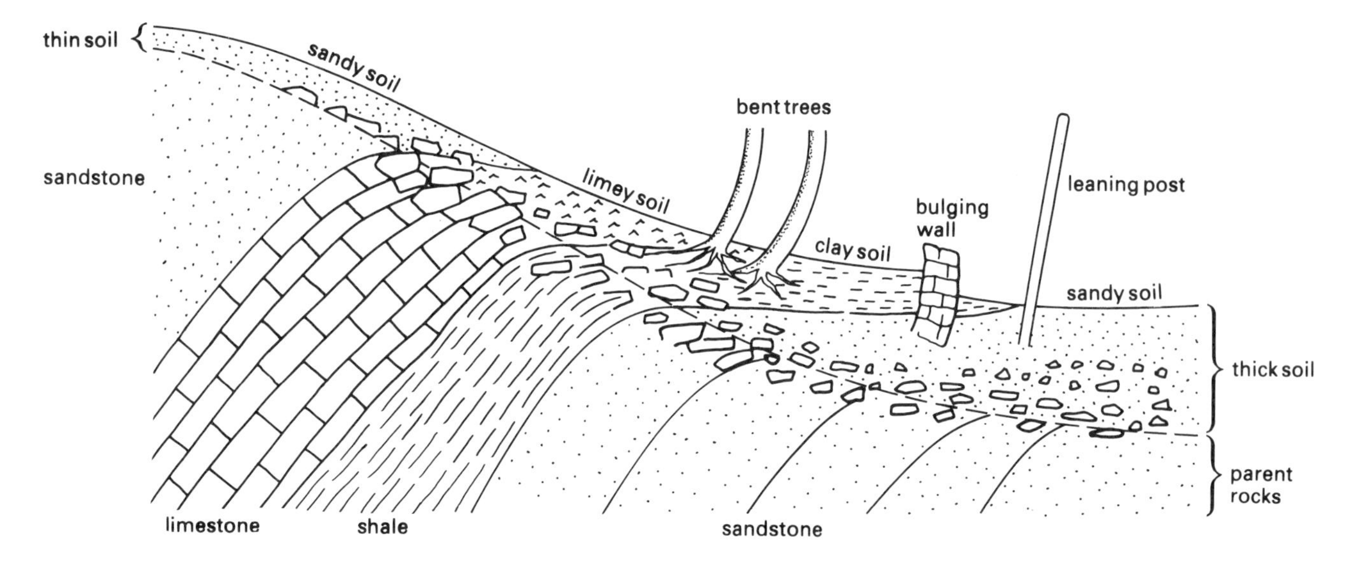

The slow movement of soil down a slope causes trees to bend as they grow, and walls to bulge. This movement means that the soil from each parent rock layer is found displaced downslope.

Fig 9.4 Soil creep

Activity 1: To show how clay flocculates when lime is added

1 Mix some clay and tap water together in a beaker. Even after some time the water remains cloudy. The fine clay stays in suspension.
2 Mix the same amount of clay with limewater in a beaker. Record what happens.
   a) How long did it take for the untreated clay to settle out?
   b) How long did it take for the clay in the limewater to settle?
   c) Could you see the clay clumping together and settling?

SOIL MOVEMENT

Soil creep

Soil often moves very slowly down a slope. This process is called *soil creep* (Fig 9.4). The process ensures that soils tend to be thicker in valleys at the foot of slopes and thinner on the hilltops.

Soil erosion

The removal of soil by water and wind erosion is a great problem, especially in dry regions where there is little natural plant cover. European farmers settling in Australia did not understand this problem. They started to clear the land and plough up the bare soil as they would have done at home. Even before the crops had a chance to grow, the dry dusty soil was picked up by the wind or washed away in sudden storms. Huge erosion gullies were carved into the topsoil after every storm. The technique of loosening the soil by ploughing only made matters worse. Plant cover can also be lost because of overgrazing by sheep and cattle. Many previously fertile areas have become deserts as the topsoil has been eroded. Once erosion starts it is difficult to stop.

Nowadays the problem of soil erosion is better understood and techniques to prevent and conserve the soil have been developed.

Soil conservation (Fig 9.5)

Some of the main methods for stopping the loss of soil are:

a) Deep ploughing below surface level, using cultivators that do not turn the soil over.
b) Planting alternate strips of grass across a field so that only small strips of soil are left bare. Any soil washed or blown away will collect against the grass strips.
c) Planting shelter-belts of trees and bushes to slow down the wind.
d) Planting the seeds of the next crop as the first crop is being harvested, so that the soil is never left bare.
e) Covering any bare patches of soil with straw to keep the soil moist and heavy.
f) Making terraces on steep slopes so that water will sink in rather than rush straight down and wash the soil away. Terraces can also help to prevent soil-creep down the slope.

INVESTIGATING SOILS

Soil is made up of five main components. The amount of each component varies in different soils. The components are:

(i) mineral matter;
(ii) water;
(iii) air;

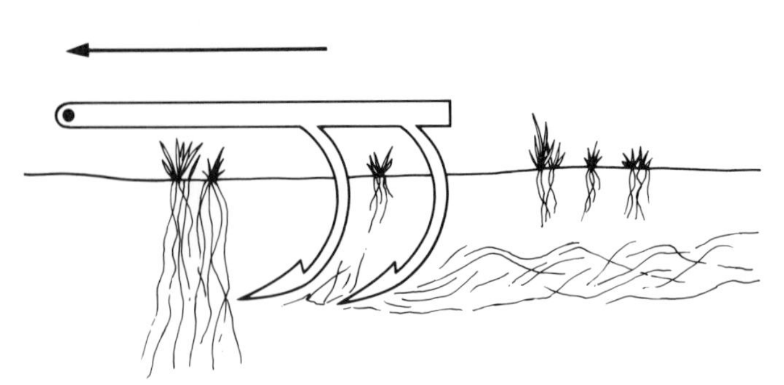

(a) A subsurface cultivator

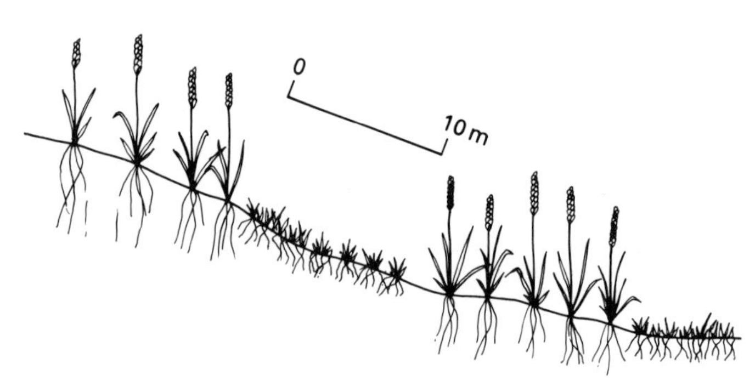

(b) Alternate strips of grass help to bind the soil

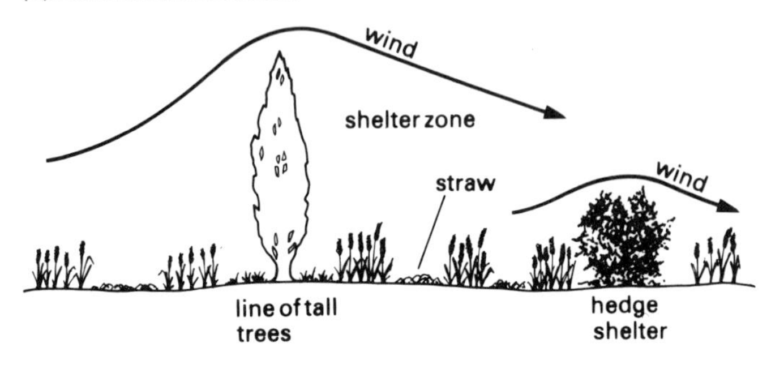

(c) Shelter belts prevent wind erosion. Crops are grown in strips. Bare soil between strips is covered with straw.

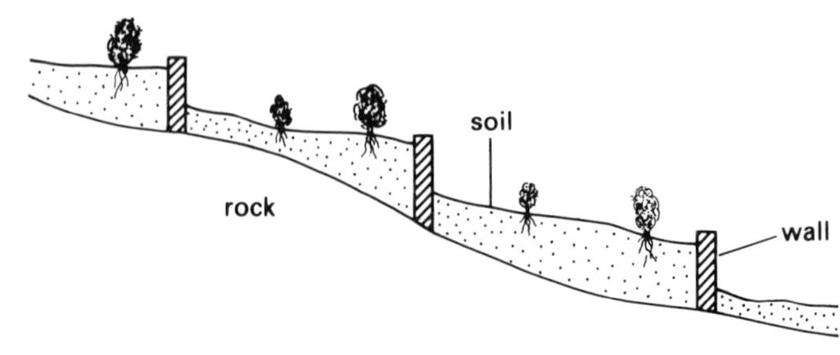

(d) Terracing of slopes helps to stop soil erosion.

Fig 9.5 Methods of soil conservation

(iv) humus;
(v) living organisms.

The activities outlined below can be used to compare the properties of different soils, and to study this variation.

Activity 2: Investigating different soils

1 Collect samples of a clay soil, a sandy soil, and a loam.
2 Examine each one on a sheet of paper.
3 Record what each soil is like on your copy of this table.

Observation	*Clay*	*Sand*	*Loam*
Colour			
Fragment size – regular or irregular?			
Texture – gritty or silky?			
Moistness			
Smell			

4 Sort out each sample into at least three piles:
Organic material a) Living things, eg, worms, insects.
  b) Dead things, eg, leaves, root fibres.
Inorganic material c) Mineral grains
What proportion of each sample is organic matter?
5 Examine the material more closely with a hand lens. Make sketches of what you see.

Activity 3: To measure the volume of air and different materials in soil

1 Put 100 cm^3 (cm^3/ml) of soil into a measuring cylinder and add 100 ml of water. The air spaces in the soil will fill up with water (Fig 9.6).
2 Work out the volume of air in the soil sample as follows:
Volume of soil = 100 ml
Volume of water = 100 ml
Soil + water = x
Therefore volume of air = 200 − x

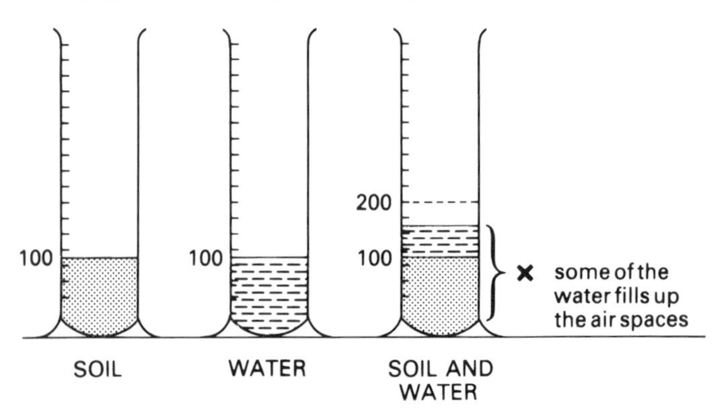

Fig 9.6 Measuring the volume of air in soil

Use this method to compare the amount of air in different soils. Which kind of soil has the greatest amount of air space – a clay, a sand, or a loam?
3 Cover the cylinder and shake it to thoroughly mix the soil and water. Then allow it to settle out. The heavier grains will settle first. The fine clay will stay in suspension much longer.
4 When the water is clear, record the volume of each layer you can see. (Use the scale on the side of the cylinder as a measure.)
5 Make a labelled sketch of the soil layers for each kind of soil tested in this way.

Activity 4: How much water is there in different soils?

1 Using a top pan balance, weigh samples of clay, sand and loam soils.
2 Allow each sample to dry out in a warm place, eg, above a radiator.
3 When the samples are thoroughly dry weigh each one again. What is the mass of the water?
4 Work out the percentage of water in each sample using this formula:

$$\frac{\text{Mass of water} \times 100}{\text{Original mass}} = \text{Percentage of water}$$

5 Record your results in your copy of this table:

Results in grams (g)	*Clay*	*Sand*	*Loam*
Original mass (a)			
Mass after drying (b)			
Mass of water = (a) – (b)			
Percentage of water			

Questions
a) Which soil contains the highest percentage of water?
b) Which soil contains the lowest percentage of water?
c) Give possible reasons for these results.

Activity 5: How much humus is there in different soils?

1 Use a top pan balance to record the mass of dry samples of clay, sand and loam.
2 Place each sample on a tin lid and roast it over a bunsen flame. This will burn off the humus.
3 Reweigh each sample. What is the mass of the humus?
4 Work out the percentage of humus in each sample using this formula:

$$\frac{\text{Mass of humus} \times 100}{\text{Mass of dry soil}} = \text{Percentage of humus}$$

5 Record your results on your copy of this table:

Results in grams (g)	*Clay*	*Sand*	*Loam*
Mass of dry soil sample (a)			
Mass of sample after roasting (b)			
Mass of humus = (a) – (b)			
Percentage of humus			

Questions
a) Which soil has the highest percentage of humus?
b) Which soil has the lowest percentage of humus?
c) Give possible reasons for these results.

Exercise 1

Pair the following phrases together to form correct sentences.

1 Soil contains decaying organic material	which help to bind the soil together.
2 The topsoil layer contains	has a large number of air spaces.
3 A coarsely-textured soil	is called soil creep.
4 Leaching is less likely to happen	by the action of micro-organisms.
5 It takes many years for an immature soil	called humus.
6 Lichens grow on bare rock	to improve its texture and drainage.
7 Humus is broken down	more plant nutrients than other horizons.
8 Lime is added to a clay soil	to develop into a mature soil.
9 Movement of material down a slope	and produce acids which dissolve the rock.
10 Soil erosion is prevented by plant roots	in a finely-textured soil.
11 Terracing of slopes helps to	loosen the soil below ground.
12 Planting shelter belts of trees helps to	are often planted in strips.
13 To prevent soil erosion, crops	stop the soil being blown away.
14 Sub-surface cultivators cut off the roots of weeds and	stop the soil being washed away.

Exercise 2

1 List under two headings the advantages and disadvantages of (a) a clay soil; (b) a sandy soil:
For example:

Sandy Soil

Advantages	*Disadvantages*
well drained	poor in mineral salts

2 A loam is intermediate in texture, between a sand and a clay.
a) Make a sketch of the texture of a loam. Base your diagram on Fig 9.3 and on the results of your experiments.
b) Describe what a loam soil is like. Base your description on the known advantages of clayey and sandy soils.

3 Ploughing is a technique often used by farmers. Below are some reasons for ploughing a clay soil. Complete each sentence using a word from the list at the end of the question.
Ploughing is carried out . . .
a) to feed the birds with upturned ________.
b) to plough into the ground weeds and unwanted crops, to provide extra ________.
c) to dry out the soil and improve ________.
d) to aerate and loosen the ________.
e) to bring compacted lumps of soil to the surface so that they will be broken down by ________.
f) to stop run-off of ________.

drainage rainwater earthworms soil humus frost-action

4 Which of the above reasons do you think is (a) the most important and (b) the least important; from a farmer's point of view? Would a conservationist list the same two reasons? If not, why not?

5 Write down the names of the five main components found in soil.

6 An experiment was made to find the water content of three different soils. (See Activity 4 for the experimental method.)
This table shows some of the results.

Results in grams (g)	*Clay*	*Sand*	*Loam*
Original mass	50	80	70
Mass after drying	40	75	60
Mass of water			
Percentage of water			

a) Copy the table and enter the figures for the mass of water for each soil.
b) Work out the percentage of water for each sample, using the formula shown in Activity 4. Enter the results on the table.
c) Which soil has the greatest percentage of water?
d) Explain fully these results.

7 The same dried soil samples were then tested for their humus content. (See Activity 5 for the method.) The results are shown in this table.

Results in grams (g)	*Clay*	*Sand*	*Loam*
Original mass	60	70	40
Mass after roasting	56	69	35
Mass of humus			
Percentage of humus			

a) Copy the table and enter the figures for the mass of humus in each soil.
b) Work out the percentage of humus for each sample, using the formula shown in Activity 5. Enter the results on the table.
c) Which soil sample contains the highest percentage of humus?
d) Why were the samples dried before the humus was burnt off?
e) What reasons can you give for this set of results?

8 Equal amounts of gravel, sand, silt and clay are shaken with water in a jar. When the mixture settles, four layers are visible.
a) Make a labelled sketch to show the most likely arrangement of the layers.
b) What kind of material would be floating at the surface?

Exercise 3

1 Explain how rocks are broken down in the weathering process to produce soil.
2 Outline the importance of plants in the formation of soil.
3 Write descriptive notes on the following:
humus micro-organisms plant nutrients earthworms
4 a) How many earthworms would you probably find in one hectare?
b) Work out the number of earthworms you would expect to find in 1 m^2.
5 a) Make a labelled sketch of a soil profile.
b) Explain how a soil profile develops, and the main characteristics of each horizon.
6 The subsoil is usually very stony. It also often has a high clay content. Where has this clay come from?
7 Why does a sandy soil often form from a granite parent-rock?
8 What rock type would you expect to find under a lime-rich soil?
9 Why is humus added to clay soils?
10 What can be added to sandy soils to improve their texture?

11 Why does the soil layer tend to be thinner on hilltops and thicker in valleys?

12 Explain how man's activities can cause extensive soil erosion.

13 Make labelled sketches and descriptive notes outlining the main methods of soil conservation.

14 a) What is meant by 'flocculation'?

b) Describe an activity that shows how flocculation occurs.

c) What is the main practical use that a farmer might make of flocculation?

10 FOSSILS

How do we know that there were once such animals as dinosaurs? How do we know that the largest dinosaur was 30 metres tall and ate plants, and that Tyrannosaurus was a flesh-eating monster? (Fig 10.1)

Dinosaurs are now extinct. In fact they died out long before humans appeared. Even our caveman ancestors never met one face to face. All that we know of them comes from a study of their fossil remains (Fig 10.2). The remains include bones, teeth, eggs, and even footprints. When the fossil skeleton of a dinosaur is put back together, we can see the size and shape that the animal must have been and work out some things about its life style. For example, fossil teeth are evidence of either a plant- or meat-eating diet; long slashing fang-like teeth are only well-developed in flesh-eating animals.

WHAT ARE FOSSILS?

Fossils are the remains or traces of once-living organisms (ie, animals and plants). They are found preserved in rock layers. A fossil may be all or part of an organism, such as a shell or a leaf; or it could be the *trace fossil* of an animal.

A trace fossil is the marking left on a rock layer, of the *activity* of an organism. Footprints and worm burrows are common examples. The fossilised droppings of animals, called *coprolites*, are also trace fossils.

How often do fossils form?

It is rare for an organism to become a fossil. Usually when a plant or animal dies, it is eaten or it rots away. Quick burial under layers of sediment may stop the organism being destroyed. This is more likely to happen on the sea bed. Even on the sea floor, a dead body is usually eaten by scavengers. Shells and bones are often destroyed by waves and currents before they can be buried. A hard-shelled animal that is buried in

Fig 10.1 A tyrannosaurus skeleton. Its overall length is 12 metres.

Fig 10.2 Dinosaur eggs – a nest of eggs found along with the remains of a ceratopsid dinosaur, in Mongolia

soft sediment has the best chance of fossilisation. Even after burial there is a high risk that the material will be destroyed. Burrowing animals could eat it; acidic water passing through the sediment could dissolve it away.

What are the chances for animals and plants that are entirely soft-bodied? They are rarely found as fossils. A few examples of fossil jellyfish have been found but they are only preserved in very special conditions.

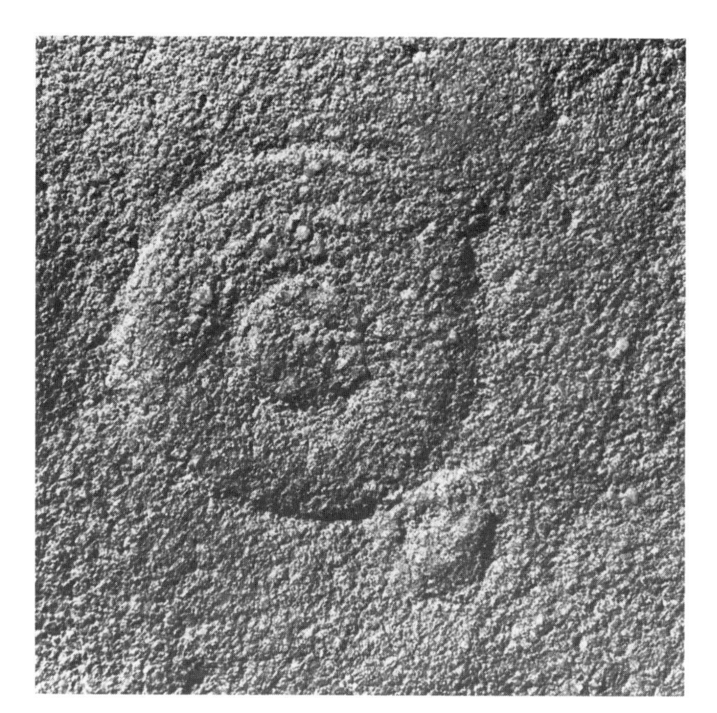

Fig 10.3 The impression of a jellyfish from Australian Precambrian rocks. This totally soft-bodied creature very rarely forms fossils. Yet jellyfish have been common sea animals for over 600 million years!

After death, the soft jellyfish falls onto fine mud and makes an impression of its shape. This is preserved when the mud changes to shale (Fig 10.3). This can only happen in very still water. The odds are even more weighted against soft-bodied land living creatures such as worms or slugs. At least jellyfish are in the right kind of environment.

The fossils found in any rock layer are therefore bound to represent just a few of the animals and plants actually living at that time. A present-day coral reef supports many hundreds of different kinds of animals and plants. Only about two per cent of organisms are likely to be represented as fossils.

Activity 1: How likely are animals and plants to form fossils?

1 Make a copy of Fig 10.4.
2 Write in the names of the animals and plants in the best positions for fossilisation, selecting from the list below. Think about:
   a) what environment the organism lives in;
   b) whether it has hard parts;
   c) those organisms with a greater chance of becoming fossilised should be written in the more densely-shaded part of the box. Four examples have already been written in for you.

slug snail earthworm butterfly caterpillar horse bird tortoise turtle crocodile fish octopus jellyfish coral burrowing shellfish seaweed moss oak tree

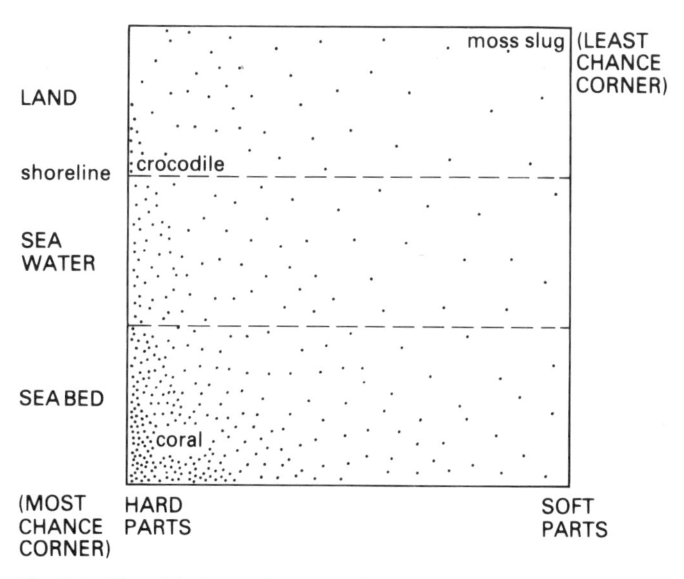

Fig 10.4 The odds 'for' and 'against' fossilation

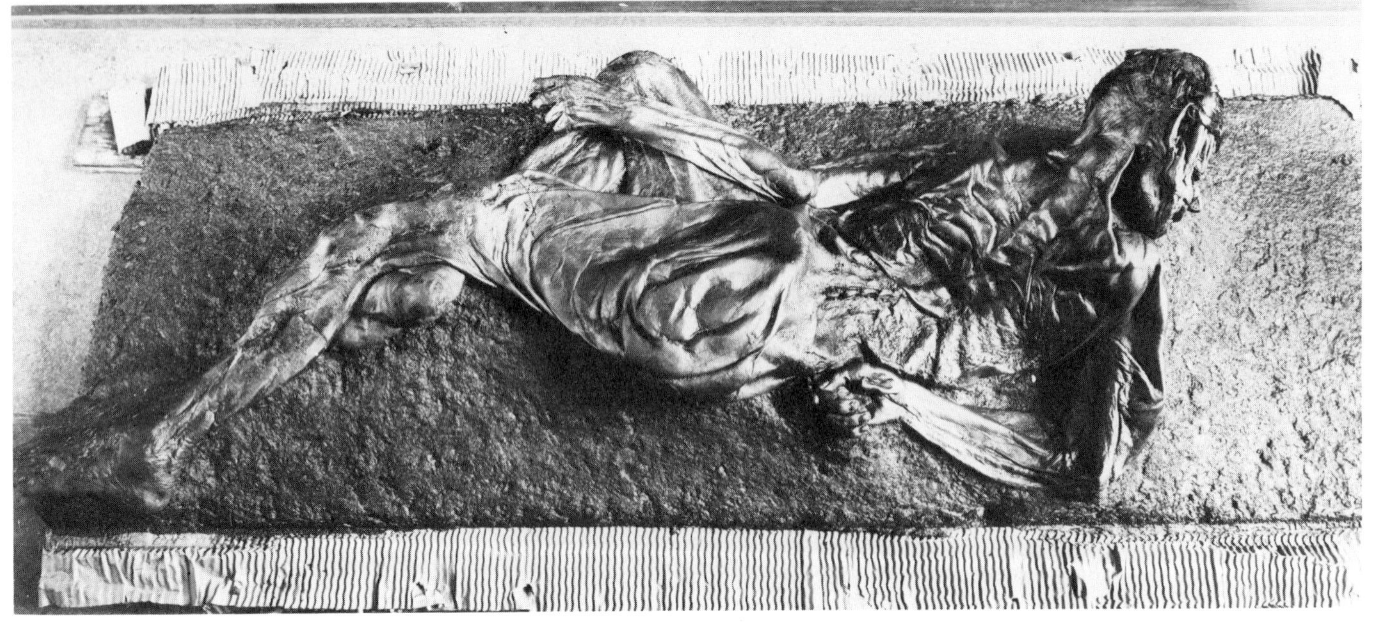

Fig 10.5 A human body preserved in Danish peat bog

KINDS OF PRESERVATION

Fossils are preserved in various ways. In some rare cases, entire organisms are preserved almost unchanged. Frozen mammoths have been found in Siberia, their flesh and bones still intact. Mammoths were large elephant-like creatures with thick fur. They had been buried in frozen mud since the last ice age. There are also examples of organisms preserved in peat bogs. Peat is antiseptic. Moulds and bacteria cannot remain active. Therefore, organic material does not rot away. Wood, bone, whole animals and even humans have been recovered (Fig 10.5).

Fig 10.6 A carboniferous plant. The delicate detail of this leaf has been preserved as a film of carbon.

Preservable organic material may undergo a change as the hard parts are slowly replaced by minerals. The minerals enter the material in solution. The fossil is literally *petrified* (turned into stone). This process is called *petrifaction*. If it happens gradually, the fine detail of the tissue is preserved. In some cases, the detail is so good that even the cell structure is preserved. Organic tissue can undergo a process of slow change known as *carbonisation*. The tissues are converted to a thin film of carbon, pressed onto the rock. Plant leaves and graptolites are often preserved in this way (Fig 10.6).

Sometimes mineral solutions remove the fossil material completely. The fossil dissolves away and only a hollow *mould* of its shape is left. This cavity may later fill up with other minerals and a copy of the fossil is formed. This is known as a *cast*.

The kind of deposit in which a fossil is buried is important. Fine mud that later hardens to shale or limestone is best for preserving good detail. Most sandstones remain too porous. Groundwater passing through the rock will dissolve the fossil away. Many fossils are lost by later earth-movements, metamorphism and erosion. They are never seen.

Activity 2: Making a fossil

1. Coat the outside of a shell with some petroleum jelly. Then press the shell into some plasticine.
2. Remove the shell.
3. You now have an impression of the shell. Pour some plaster of Paris into the mould. When it sets you can pull the plasticine away and there should be a good cast of the shell.

FINDING FOSSILS

A good way to learn about fossils is to collect them. Use a geological hammer and a chisel and protect your eyes with safety glasses. The best places to look for specimens are at natural exposures of sedimentary rocks. Coastal cliffs and river banks are ideal places, especially after winter storms. Fossils may have been eroded out. Remember that limestone and shale are the best kinds of rock; be careful of loose overhanging cliffs. There are many guides on the market which will help you to identify what you have found. When you get back to school or home your fossils can be cleaned up with a fine chisel. It is a good idea to protect the cleaned specimens with a coat of clear matt varnish. Finally label your specimens and store them carefully. Cardboard boxes divided into sections are ideal for beginning a collection.

FINDING FOSSILS IN AFRICA – THE STORY OF LUCY

This story is about the search for *hominid* (human-like) fossils, in Africa. As with many land animals, fossils of our ancestors are very rare and usually only a small part of the skeleton is found. Because the fossils are so rare, finding them can be very exciting.

In 1974 an American called Don Johanson went to look for hominid fossils at Hadar. Hadar is in the centre of the Afar desert in north-east Africa. It is an ancient lake bed that has dried up. The water has gone, but the layers of sediments are still there and they record the history of past events. You can trace old volcanic ash falls and see deposits of mud washed down from distant mountains. In places deep gullies have been cut into the soft lake bed layers by new streams. It hardly ever rains in this region but when it does the water rushes down the gullies. Then the sediment is cut away and new fossils become exposed.

This is Don's story of what happened on 30 November 1974:

> Gray and I parked the Land Rover on the slope of one of those gullies. We got out and began surveying, walking slowly about, looking for exposed fossils.
>
> Some people are good at finding fossils. Others are hopelessly bad at it. It's a matter of practice, of training your eye to see what you need to see. . . .
>
> The gully had been thoroughly checked before. Nothing had been found. I decided to make that final detour. There was virtually no bone in the gully. But as we turned to leave, I noticed something lying on the ground partway up the slope. . . .
>
> It was the back of a skull. Near to it was part of a leg bone. We stood up and began to see other bits of bone on the slope – all of them hominid. They were all parts of a single skeleton. No such skeleton had ever been found anywhere. . . .
>
> That night we never went to bed at all. We talked and talked. There was a tape recorder in the camp, and a tape of the Beatles' song 'Lucy in the Sky with Diamonds'. At some point during that evening – I no longer remember exactly when – the new fossil picked up the name of Lucy.

What is Lucy?

Johanson had found the partial skeleton of a female hominid. Although an adult she was only just over one metre tall. The fossil is 3.4 million years old, and is the oldest best-preserved fossil skeleton of any erect walking human ancestor ever found (Fig 10.7). Footprints of animals similar to Lucy have been found at another African site called Laetoli. They were preserved in ash deposits of about the same age and clearly showed that hominids could walk erect.

Because so few fossils have been found it is difficult to know how these early fossils relate to human beings. They were not true human beings. It is more correct to think of them as near relatives. They were also probably extinct by the time true humans appeared about 100 000 years ago.

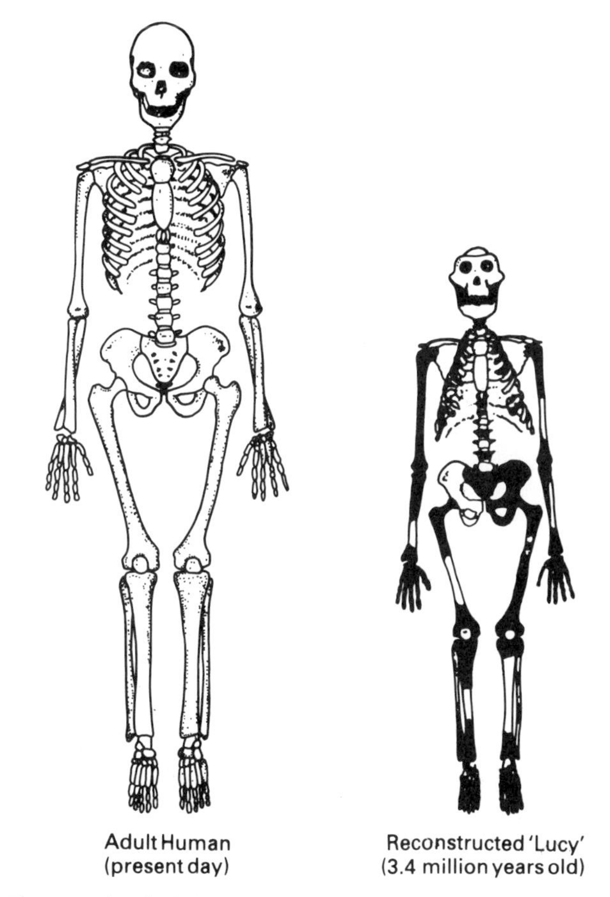

Fig 10.7 The shaded parts of the 'Lucy' skeleton represent fossil bones actually found

FOSSIL GROUPS

When you begin to collect fossils you will need to find out what they are. The descriptions that follow should help you to identify most of the fossils you find. (For more detailed identification, refer to the books quoted in the Bibliography, page 145). Exhibits of fossils in museums are also useful: you can compare your fossil with those on display and find out what it is. As you read the descriptions, look at Fig 10.8, it tells you the life span of each of the fossil groups.

Vertebrates

Vertebrates include all animals with backbones such as mammals, birds, reptiles, amphibians and fish. They are quite rare as fossils. The skeletons often break up and are destroyed.

Invertebrates

Animals in this group do not have backbones. The group includes corals, brachiopods, molluscs, trilobites, echinoderms (sea-urchins, sea-lilies), and the graptolites. Invertebrates are more easily fossilised than vertebrates and are found in many sedimentary rocks.

Corals
Corals belong to the same group of soft-bodied animals as jellyfish and sea-anemones. The group name is *Coelenterates*. Each coral polyp produces a skeletal stem made of calcite, known as a *corallite* (Fig 10.9).

Corals may live in colonial reefs or in solitary form. Reef corals need tropical, shallow, well-lit waters to survive. Fossil reefs formed at different times in the geological past, especially in the Palaeozoic era.

Brachiopods
Brachiopods have shells consisting of two hinged valves, which open for feeding. The animal is fixed to the sea floor by *a pedicle* (a fleshy stalk). In many species, this passes out through an opening in the larger valve (Fig 10.10).

Over 30 000 different fossil and living brachiopod species have been identified. They were especially important in the Palaeozoic and Mesozoic eras. They are still quite common, with 200 known living species in modern seas.

Molluscs
This large and varied group of animals includes the snails, cockles and cuttlefish living today, as well as the important fossil ammonites and belemnites. The group divides into three main sub-groups: the bivalves, cephalopods and gastropods.

Bivalves
Living members include cockles, mussels, scallops, oysters and clams. Like the brachiopods, they consist of two valves. At first sight, bivalves look very similar

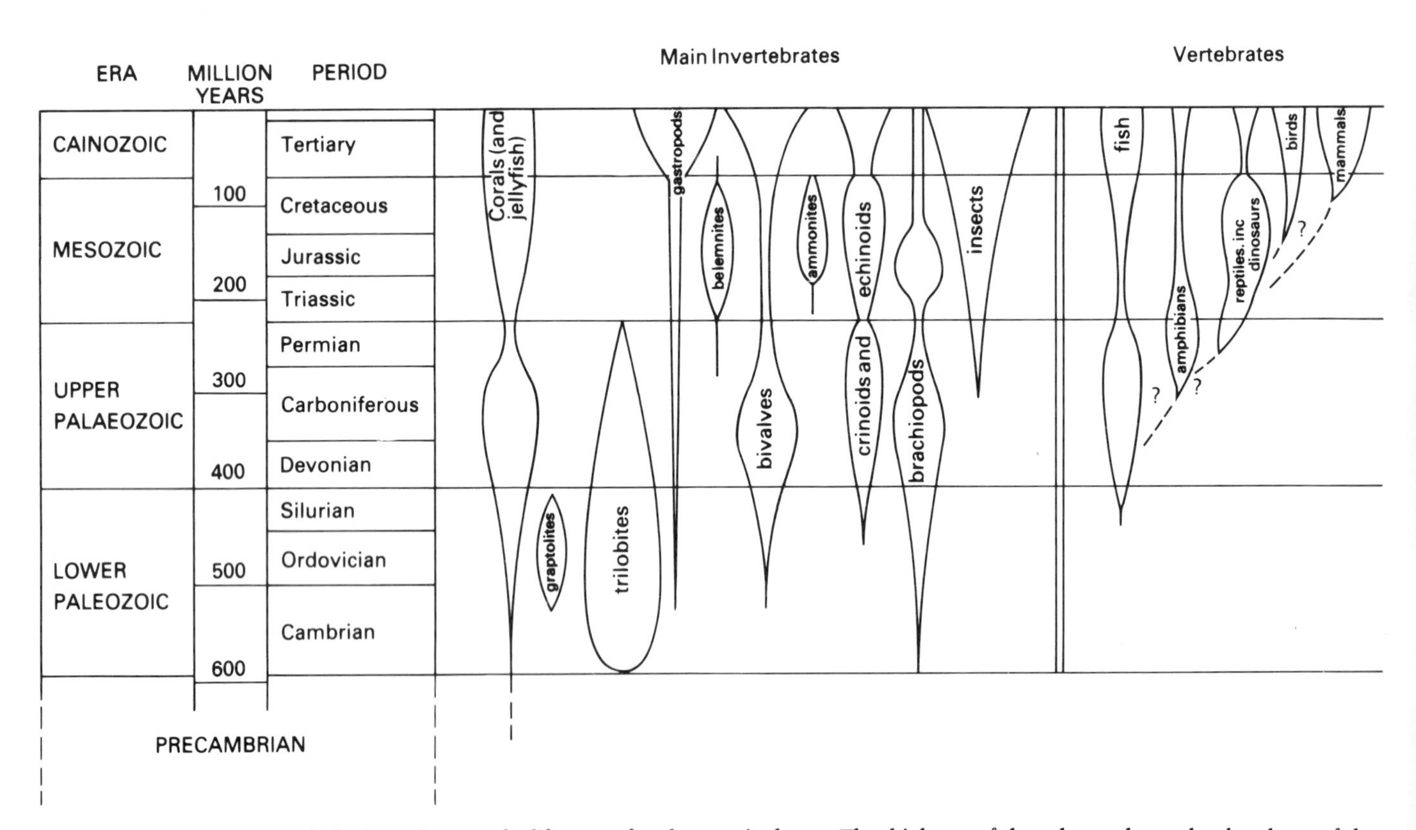

Fig 10.8 The development of life through time. The life-span of each group is shown. The thickness of the column shows the abundance of the group.

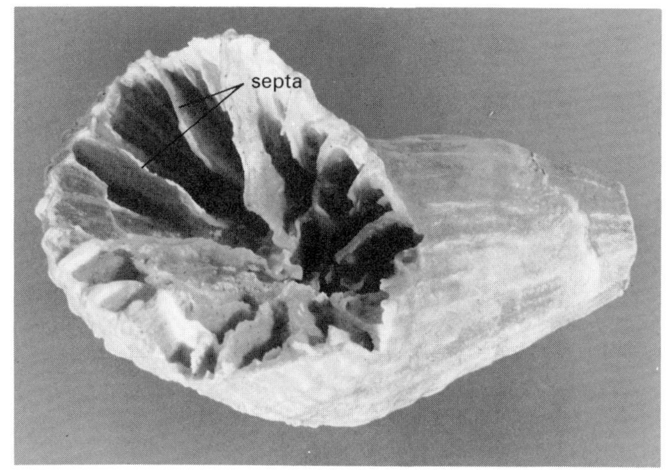

Fig 10.9 Two species of coral from the Carboniferous limestone
(a) The solitary *Parasmilia* (b) *Lithostrotion*, an example of a colonial coral. Note the vertical septa in both examples.

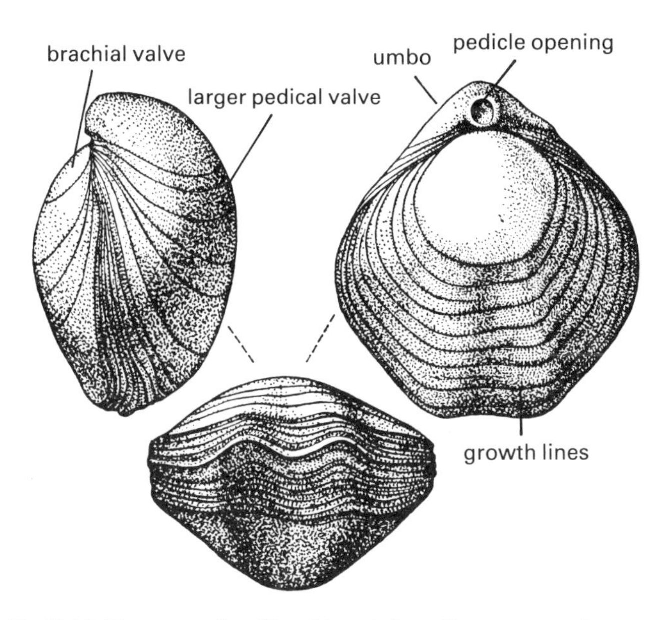

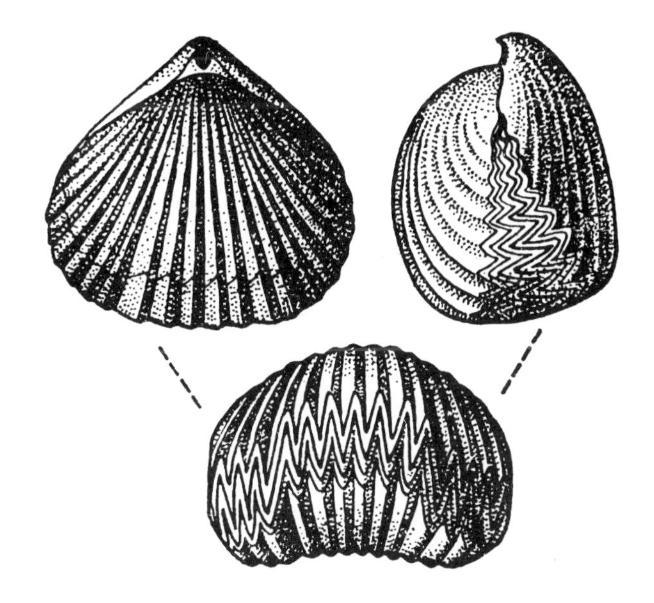

Fig 10.10 Two examples of brachiopods from Cretaceous rocks

to brachiopods but there are ways to tell them apart:

(i) Brachiopods usually have one valve that is larger than the other, whereas in bivalves they are usually the same size.

(ii) Each valve of a brachiopod can be divided along its mid-line and the two halves are then mirror images. In other words, the valves are *bilaterally symmetrical*. Bivalves do not have this symmetry (Fig 10.11).

Bivalves are very common Mesozoic and Cainozoic fossils, and are still abundant in present-day seas.

Cephalopods

This entirely marine group includes the living squid, cuttlefish and octopus. The two most important fossil cephalopods are the belemnites and ammonites. Both these groups are now extinct.

(i) Belemnites were shaped rather like bullets (Fig 10.12). The main skeletal part of the body was

Fig 10.11 Bivalve from rock of Tertiary age

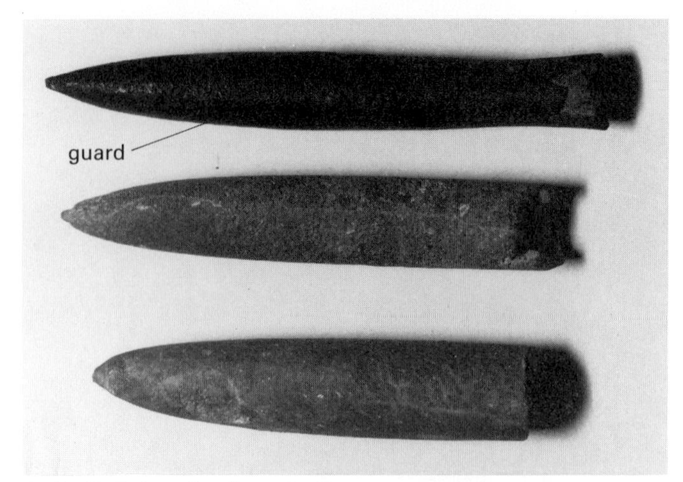

Fig 10.12 Belemnites from the Cretaceous period

Fig 10.13 *Harpoceras* ammonite – shelly limestone, from rocks of Jurassic age

Fig 10.14 Gastropod (*Athleta*) from rocks of Tertiary age

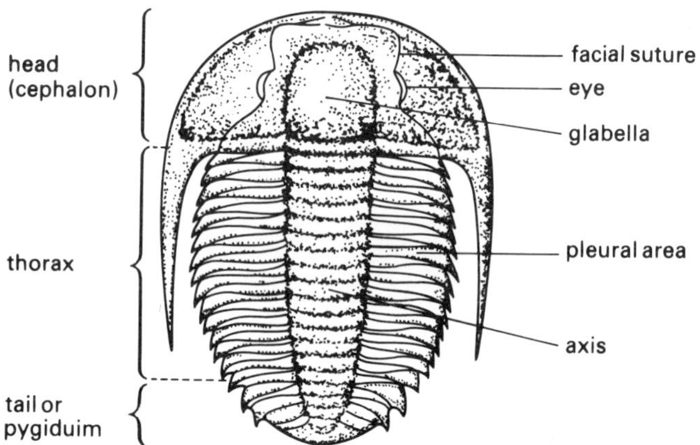

Fig 10.15 (a) Trilobite from rocks of Ordovician age (b) Diagram of *Angelina* trilobite

the *guard*. In front of the guard was a cone-shaped shell that contained some of the soft parts of the animal. Belemnites were squid-like animals with long tentacles.

They are especially common in rocks of Mesozoic age.

(ii) Most ammonites had coiled, chambered shells that displayed a great variety of ornamentation and size (Fig 10.13). Their fossils are found only in rocks of Mesozoic age. During this time they were very successful.

The nearest living relative to the ammonite is Nautilus. This animal also has a shell of many separate chambers. It lives in the last outer chamber. The other chambers are filled with gas and give the animal buoyancy. Structural comparisons of Nautilus to ammonites suggest that the extinct ammonites had a very similar free-swimming mode of life.

Gastropods
Most gastropods have a single spirally-coiled shell. The group includes whelks, winkles, land slugs and snails. Marine snails are the commonest fossils. Nearly all gastropods are crawlers. They are found on the sea-bed in freshwater and on land (Fig 10.14).

Trilobites
Trilobites are a group of extinct *arthropods.* Arthropods have an outer skeleton of jointed segments. The group includes present-day crabs, lobsters and shrimps, as well as spiders and insects. Arthropods are the most abundant invertebrates; there are over one million different living species.

Trilobites are only found in Palaeozoic rocks. The body was divided into three lobes (parts) by a pair of furrows running the full length of the body: the central lobe, the *axis*; the two side parts, the *pleural lobes* (Fig 10.15). At intervals during growth, trilobites shed their skeleton. The *facial sutures* split open, and the animal crawled out.

Echinoderms
Another group of marine animals is the echinoderms. It includes the starfish, sea-cucumber, sea-lily and sea-urchin amongst the 5000 different living species.

Crinoids ('sea-lilies') and *echinoids* ('sea-urchins') are the most important echinoderm fossils. They have a *test* (shell) of calcite plates that are interlocked and held between two layers of soft tissue.

(i) Crinoids are animals but they look very like plants and are usually fixed to the sea floor. They have a small body at the top of a stem (Fig 10.16). The mouth faces upwards and is surrounded by many food-catching arms, that may have a fern-like appearance. The stem is built of small disc-like plates called *ossicles.* Crinoids usually have fragile skeletons. It is rare to find complete fossil specimens. However, the stem plates are easy to find and may form crinoidal limestones.

(ii) Echinoids have spherical bodies, sometimes covered in long protective spines. The mouth is downward facing. The anus is usually towards the top of the test. Echinoids are commonly found in many Mesozoic and Cainozoic rocks. The group divides into two major sub-groups: regular echinoids and irregular echinoids (Figs 10.17 and 10.18).

All echinoids have a crawling mode of life, feeding on plants or other animals.

Fig 10.16 A Silurian crinoid

Fig 10.17 A *Micraster* echinoid

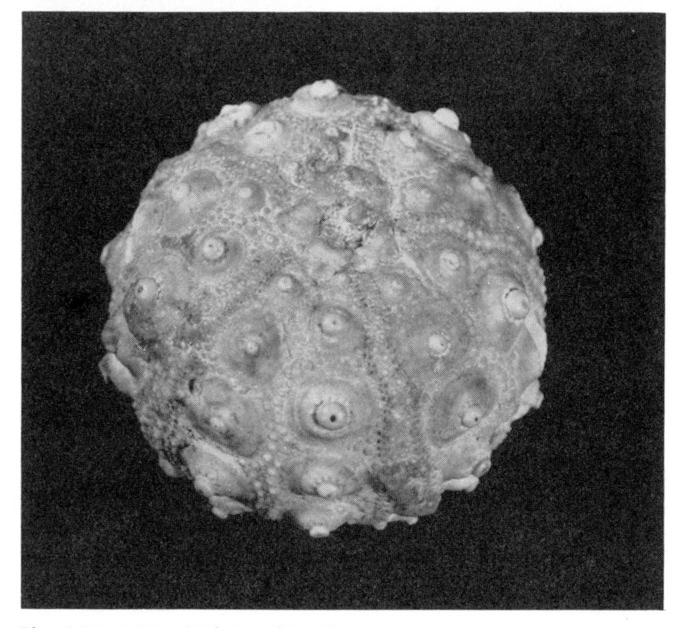

Fig 10.18 A *Hemicidaris* echinoid

Graptolites
Many of the bedding planes of early Palaeozoic sediments are marked by very thin pencil-like streaks of carbon film (Fig 10.19). These are extinct graptolites. They were small colonial animals, living as part of the floating plankton. Closer inspection of the fossils reveals a fret-saw like appearance to each branch of the colony. The individual 'frets' are known as *cups* or *thecae*. Each one housed an individual graptolite.

Graptolites reveal a distinct pattern of development through time. Early graptolites had many *stipes* (branches). As time went by, there was a trend to fewer branches (Fig 10.20). Most early graptolites were *pendent*, the branches hanging down. However the graptolites of the Silurian period were *scandent* (curved back) the stipes were joined back to back and the cups faced outwards.

The change and development of graptolites over several million years is a good example of *evolution*.

EVOLUTION

Evolution is the theory that living things can adapt and change. They evolve in response to changes in their environment. Every kind of animal or plant will show *variation*. For instance, there are millions of individual human beings. Each one of us is slightly different. Our fingerprints vary, so does the colour of our eyes. Evolution happens when one particular character is *selected*. For instance, a change to a cooler climate could favour those animals with more hair. Those with less hair would be less able to survive. This could explain why animals like the woolly mammoth and woolly rhinoceros appeared. They evolved because of *natural selection*. Dog breeders select varying characters more deliberately. This is a form of artificial evolution. They breed animals that have similar features. Slowly, over many generations, the differences between for example, alsatians and lap-poodles, become more pronounced. Both breeds are dogs, however.

It is important to realise that evolution is a very slow process. It may take 10 thousand years or longer for a distinct change to occur in a group of organisms. Therefore, it is difficult to prove exactly how evolution takes place. Fossils can be used as evidence of the main trends in the evolution of a particular group of organisms, as we have seen in the case of graptolites. Another good example of this is seen in the sea urchin *Micraster*. It evolved throughout the time the Upper

Fig 10.19 Didymograptus 'tuning fork' graptolites from rocks of Ordovican age

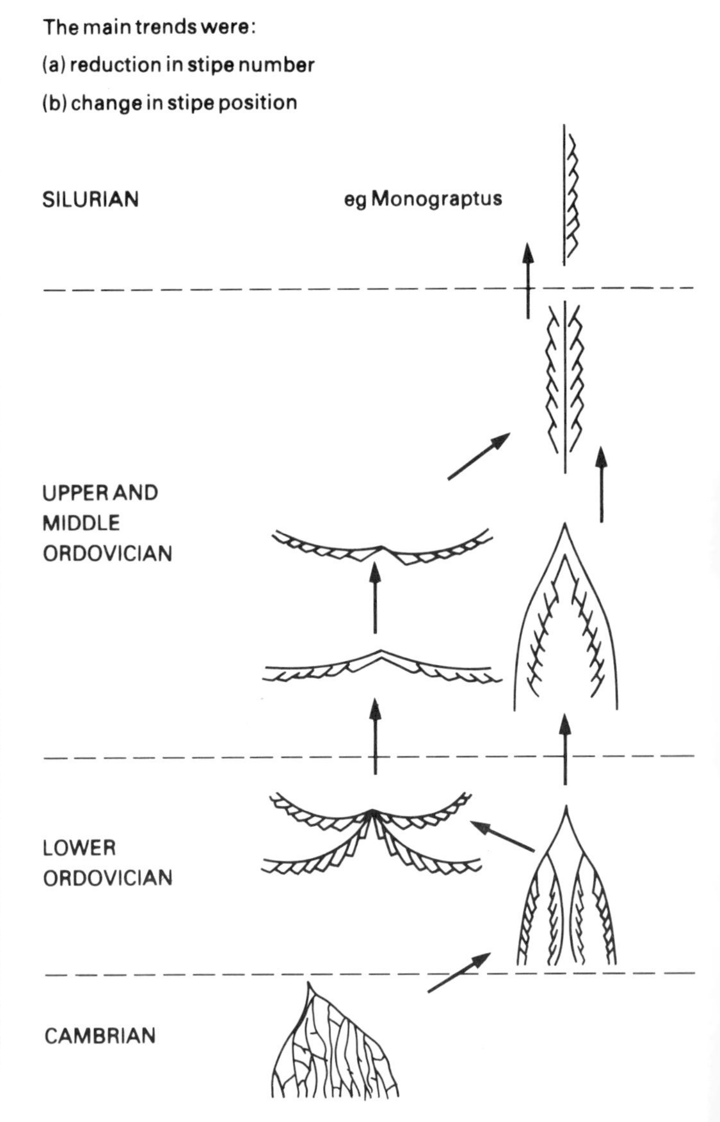

Fig 10.20 The evolution of graptolites

Chalk beds of Cretaceous age were being deposited – a period of about 10 million years. During this time there was an increase in size and height of the test and a deepening of the grooves on the top of the test.

Fossils like Micraster and the graptolites are very useful in telling us the *age* of the rocks in which they are found. For example, the presence of Monograptus in a rock layer means the rock is Silurian in age (Fig 10.20).

THE HISTORY OF LIFE

The Precambrian era

No one is exactly sure when life first appeared on Earth. The earliest signs of life in the fossil record are found in rocks over 3000 million years old. Such fossils are extremely rare and difficult to assign to any particular group of organisms. There was definitely marine plant life 2600 million years ago. Fossils of simple algae have been found in Precambrian limestones.

One of the oldest fossils in the world was found by a schoolboy. It is a sea-pen called *Charnia* (Fig 10.21). Sea-pens are soft bodied coral-like animals. The fossil was found in rocks of Precambrian age at Charnwood Forest, Leicestershire. The Charnia fossil is 690 million years old. Rare fossils of jellyfish and worm-like animals of similar age have also been found.

Fig 10.21 The Precambrian *Charnia* fossil

The Lower Palaeozoic era
(Cambrian, Ordovician, Silurian)

The era began 570 million years ago. At this time there was a sudden increase in the number of fossils. This is because many invertebrates evolved skeletons. This was one of the major events in the history of life.

For most of the era, all life was confined to the sea. Vertebrates (animals with backbones) appeared towards the end of the era. These vertebrates were the armoured and biting fish. By the end of the era, the earliest land plants had appeared. They began to grow mostly in shoreline swampy areas.

The Upper Palaeozoic era
(Devonian, Carboniferous, Permian)

During this era thick forests grew on many land areas. The remains of these forests now form important deposits of coal throughout the world. Insects and spiders successfully colonised the land. Amphibians and reptiles were other notable newcomers. As the era ended, many sea-living invertebrates became extinct and there was a great reduction in the variety of sea-life. Trilobites died out completely, as did many kinds of corals, brachiopods and crinoids.

It is difficult to explain why so many successful groups of animals should all die out at about the same time. One possible explanation is that due to plate movements, the continents came together to form a single supercontinent (see Chapter 3). This change resulted in the loss of a great variety of shallow water environments, the habitats for these animals.

Large areas of interior desert were created in the Permian period because of the changes in climate caused by the formation of a supercontinent.

The Mesozoic era (Triassic, Jurassic, Cretaceous)

Early in this era, the supercontinent began to break up and the pieces drifted apart. This created a whole variety of new environments which were colonised by new species and new kinds of plants and animals. Some new kinds of corals and brachiopods developed and ammonites, belemnites and gastropods became very common. There were many more bivalves, eg, the oyster, and the molluscs emerged as the most successful invertebrates. Sea-living vertebrates included sharks and many reptiles.

Life on land was also very varied. The first grasses and modern trees appeared. The climate was much warmer than today. This favoured the development of reptiles. The Mesozoic is most notable for the land-living dinosaurs. This varied group of reptiles include the massive *Diplodocus* with a skeleton over 26 metres long and the little *Compsognathus*, the size of a chicken. Different forms of flying reptiles also developed. The first bird, *Archaeopteryx*, also dates from this time (Fig 10.22). Its skeleton is very like that of a small dinosaur but it had feathers and could fly. Unlike

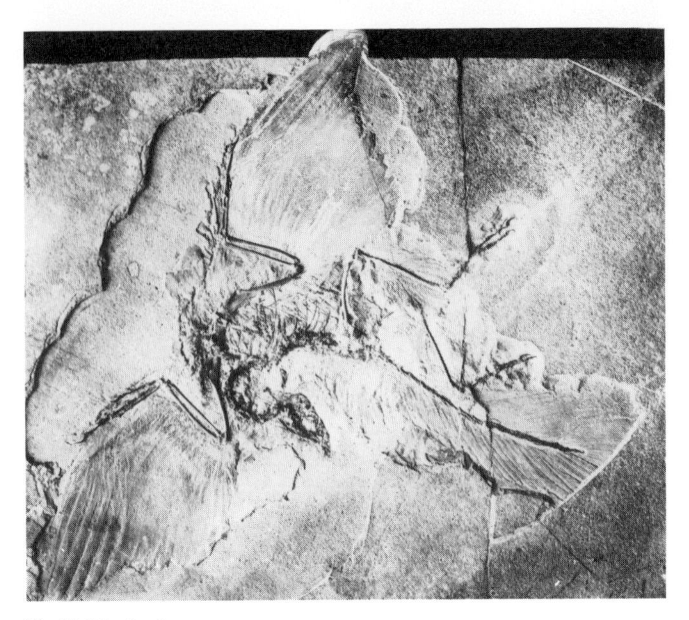

Fig 10.22 *Archaeopteryx*

birds today, it had teeth. Some rare fossils of early mammals have also been found.

Towards the end of the Mesozoic era, about 63 million years ago, the dinosaurs, flying reptiles and most of the sea-living reptiles suddenly became extinct. It is strange that this should happen when these animals were apparently at the peak of their development. Was the cause some kind of disease perhaps, or a cooling of the climate? Reptiles are especially sensitive to temperature change because they cannot keep their body temperature constant. They need constant warmth from the sun to function properly.

Reptiles were not the only casualties. Many kinds of sea-living invertebrates disappeared, including the ammonites. Many forms of surface plankton also died off. Why did this happen? A valuable clue is provided by the *iridium clay layer*. This is a very thin deposit found in different parts of the world and it dates from the end of the Mesozoic era. This distinct layer contains a number of rare substances, eg, iridium, platinum and gold. Where did the clay come from? Could it be the settled dust from a large volcanic explosion? Or did a large asteroid from outer space collide with the Earth?

Huge clouds of dust in the atmosphere would have blocked out the sun's heat. For perhaps a decade or so the climate would have become colder and darker. Many plants would die. So would the animals that used them for food. Being especially sensitive to temperature most reptiles would not be able to survive.

The Cainozoic era (Tertiary, Recent)

This was the golden age for mammals and birds. As the dust settled after the catastrophe, the survivors quickly colonised the new habitats open to them once the dinosaurs had disappeared. Many new groups of land animals appeared and thrived. There was plenty of food. Trees and flowering plants were common and grasslands developed in nearly every continent. Within a few million years mammals successfully managed to colonise every kind of environment. This is proved by the fact that their remains are found in every environment from the hot deserts to the polar ice caps. Many of the early mammals were huge beasts, up to four metres tall. Baluchitherum was the largest known land mammal. It was six metres tall and browsed on the top branches of trees. The flesh-eating Phororhacas was a giant ostrich-like bird. Its skull alone was the size of a horse's head. Many birds and mammals adapted to life in water, examples are the penguin, whale, dolphin and seal. Flying mammals, bats, even conquered the air.

Then in the late Cainozoic, two million years ago, came the Ice Ages. There was a succession of cold periods called *glacials*, interrupted by warmer *interglacials*. Huge ice sheets covered much of the land surface in the glacial periods. Many animals could not cope with the changes of climate and became extinct. Human beings however, developed rapidly at this time, and today their impact on other forms of life is very great. Many kinds of animals and plants are becoming extinct because of human activities.

FOSSILS AND TIME

Fossils tell us about the history of life and its development. They are also useful in dating rocks. Layers of sedimentary rocks are laid down in time order. The oldest layers are underneath progressively younger layers. In the case of the Upper Chalk it is possible to identify three different age layers (*zones*) because they each contain different Micraster fossils.

One of the first people to realise how fossils could help to sort out this time order was William Smith. He was a canal-engineer in the late eighteenth century. He noticed that rock layers were arranged in the same order wherever he saw them. He also noticed that each layer had its own *zonal* (marker) fossils. He learnt to *correlate* (match up) rock layers in different places by means of their fossils. Although a rock layer of the same age might change its nature from limestone to sandstone from one place to another; its fossils do not change. This is because, being of the same age, limestone and sandstone would both contain some of the same fossils. Fossils can be used to date and identify a particular rock layer, relative to other layers. This technique is known as *relative dating*.

This method does not give an actual age in years. However, it is possible to use radioactivity to date rocks in years. This method relies on the fact that a radioactive element *decays* (changes) at a constant rate

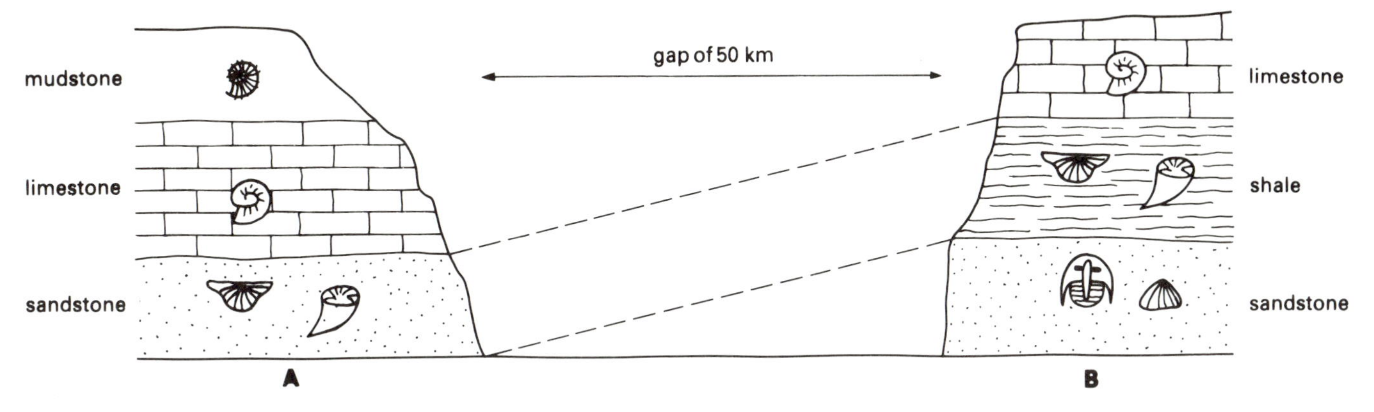

Fig 10.23 Correlation by fossils

regardless of temperature or pressure. When certain minerals are formed, they 'trap' small amounts of radioactive elements, which then start to decay. It is possible to measure proportions of the decayed amounts, against the original undecayed element. This gives the length of time that has elapsed since the rock was formed. The technique is known as *radiometric dating*. It works best on igneous and metamorphic rocks. False (older) dates may be produced by minerals in sedimentary rocks because the 'mineral grains' will have come from the breakdown of previous igneous or metamorphic rocks. Relative dating using fossils is still the most effective way of dating most sedimentary rocks.

Correlation

William Smith found that without fossil evidence it was difficult to correlate rock layers in different places. This difficulty is illustrated in Fig 10.23.

At first sight it looks as though it might be correct to match up the two sandstones at A and B, but the marker fossils in each rock layer are different. The correct way to correlate is shown by the dotted lines. The fossils indicate that the sandstone at A is the same age as the shale at B.

Not all fossils are useful for the dating and correlation of rock layers. One which is not useful is the brachiopod *Lingula*. Its fossilised shells have been found from the Cambrian up to the present day but Lingula is of little use because it has not changed its shell form very much in 570 million years! The ideal zonal fossil needs to be a creature that exists only for a short time, but spreads rapidly over a large area. Therefore rapidly evolving, free swimming or floating creatures provide the best zonal fossils. The ammonites of the Mesozoic are a good example. All ammonites were sea living creatures. They had a single-chambered buoyant shell which allowed them to float or swim over large areas. They evolved rapidly, changing their shape and size. Most ammonites had a short time range and extinct ammonites were replaced by later ones.

Fossils and rock structure

The matching of rocks using fossils is often necessary and crucial. It is done in order to work out the structure of the rocks in an area. This is important in the search for oil or coal. During exploratory drilling for oil, tiny *micro-fossils* are brought to the surface by the drilling mud. Close study of them can show the age and structure of rock layers. The depth at which the oil is likely to be found can then be forecast (Fig 10.24).

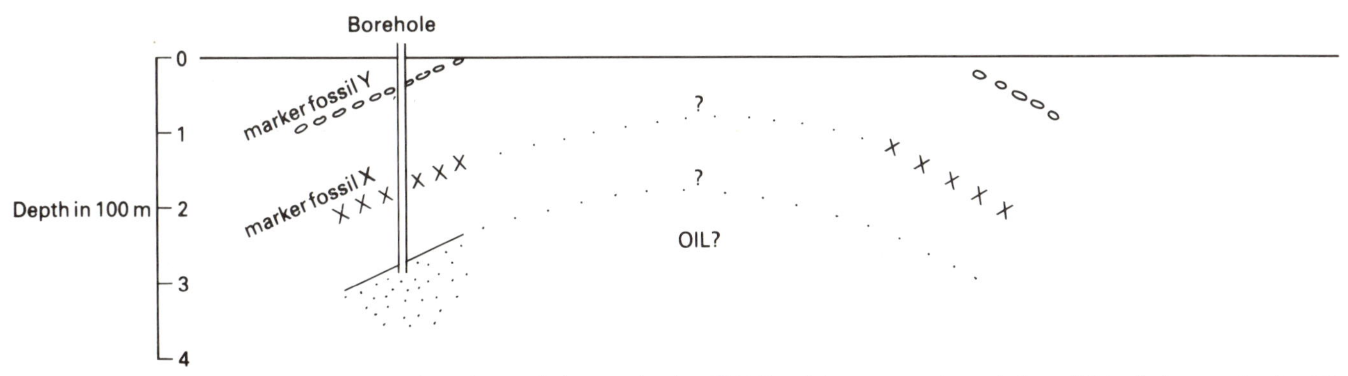

A borehole is drilled first. Marker fossil X is found 130 m below marker fossil Y. The oil-bearing sandstone is found 100 m below marker fossil X. At this point oilmen suspect that oil is trapped in an anticline structure.

Fig 10.24 Using fossils to interpret rock structures

Exercise 1: Dinosaur digestion

Rex is a keen fossil collector! Inside him are the names of at least 10 different kinds of fossil. Can you find them? They may be horizontal, vertical, diagonal or even backwards.

eggs fish teeth snail shell bird bone crab leaf coral

(*Solution on page 143*)

Exercise 2

1 There is something wrong with each of the statements below. Write out each statement in the correct form.
   a) Long slashing teeth are only found in plant-eating animals.
   b) Fossils are the remains of the hard parts of living organisms.
   c) Fossils will always form if there is quick burial after death.
   d) Fossils are never found in porous sandstones.
   e) Fine sandstone and limestone are the best rocks for preserving good detail in a fossil.

2 Complete each statement with the most appropriate phrase.
   a) Petrifaction is
      A a kind of mineral oil.
      B a fossilised bird.
      C replacement of organic tissues by minerals.
      D removal of minerals from organic tissue.
   b) A fossil mould is
      A a fossil that rots away.
      B the impression of a badly preserved fossil.
      C a kind of fossilised fungus.
      D the hollow impression of the shape of a fossil.
   c) A trace fossil is
      A the mark of the past activity of an organism.
      B a very faint impression of a fossil.
      C a rare impression of soft tissue on the rock surface.
      D a man-made copy of a fossil.
   d) Lingula is
      A an example of an ideal fossil for dating rocks.
      B an extinct kind of brachiopod.
      C a brachiopod found as a fossil in many different ages of rocks.
      D a rapidly evolving, free swimming brachiopod.
   e) Correlation is
      A a collection of microfossils from an oil well.
      B forecasting the depth at which oil is likely to be found.
      C matching up rock layers of the same age.
      D using fossils to date different rock layers.

Exercise 3

1 What is a fossil?
2 What are the conditions needed for the formation of fossils?
3 What things can happen to destroy fossil material even after it has been preserved?
4 Explain what is meant by 'petrification'.
5 Look at the leaf shown in Fig 10.6. Explain how this fossil was preserved.
6 Explain, giving examples, what is meant by a 'trace fossil'.
7 Why does there appear to have been a sudden increase in the variety of life at the start of the Lower Palaeozoic?
8 a) What is the theory of evolution of life?
   b) Explain how a study of fossils can help to provide evidence to support this theory.
9 Explain fully what is meant by a 'good zonal fossil' and give examples.
10 In the correlation exercise, Fig 10.25 there are three rock sequences several kilometres apart. Some of the rock types are different but they have the same fossils. They are therefore equivalent in age. Can you match them up correctly using their fossils?
   a) Make a copy of the diagram. Then label the oldest rock layer 'a', the next oldest 'b', and so on, up to the youngest.
   b) Draw dotted lines to connect rock layers of the same age.

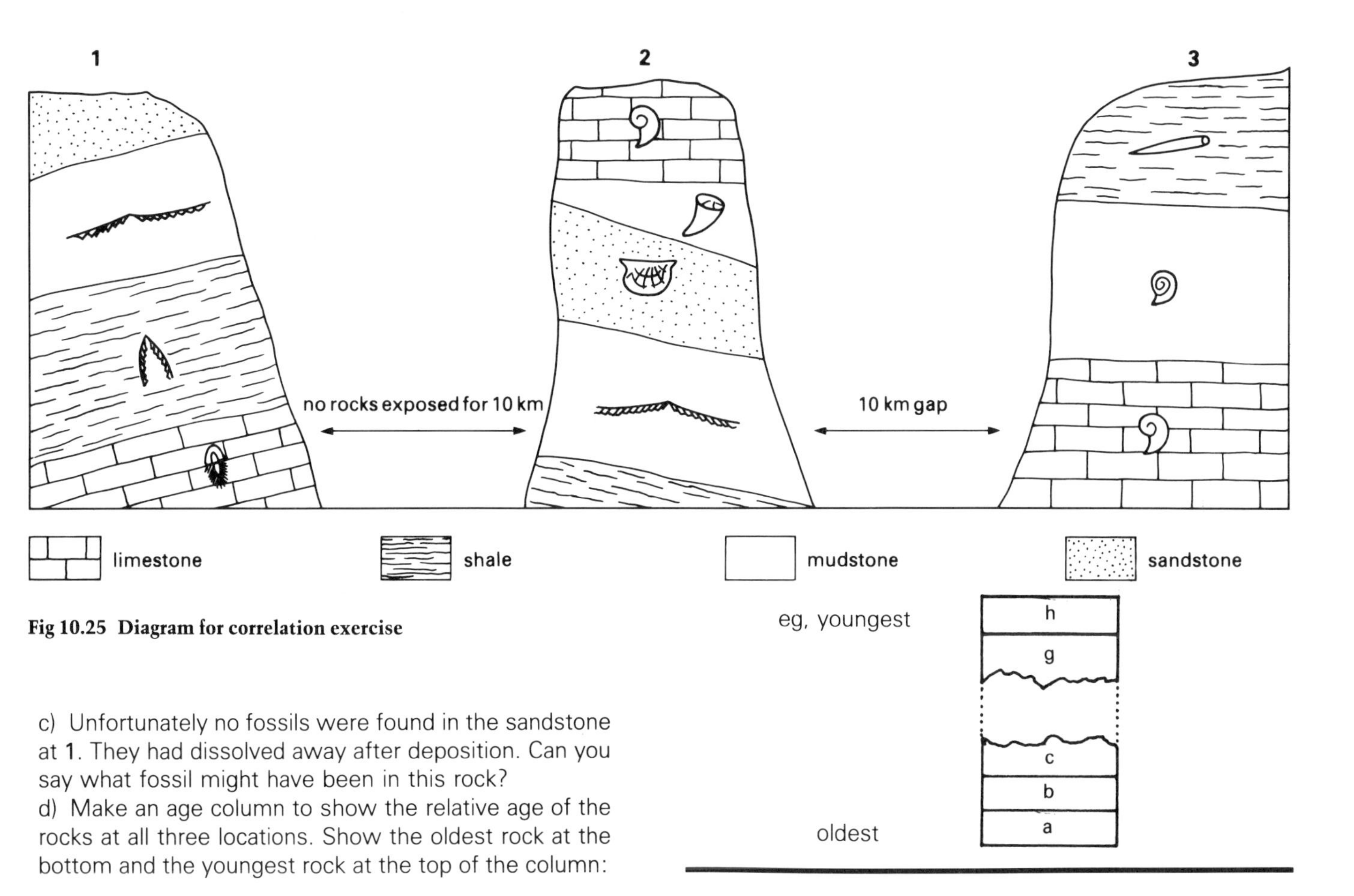

Fig 10.25 Diagram for correlation exercise

c) Unfortunately no fossils were found in the sandstone at **1**. They had dissolved away after deposition. Can you say what fossil might have been in this rock?

d) Make an age column to show the relative age of the rocks at all three locations. Show the oldest rock at the bottom and the youngest rock at the top of the column:

11 THE SEARCH FOR WEALTH

In Chapter 7, we saw how useful some minerals can be. Such minerals are known as *ores*. It is important to be able to find these minerals so they can be mined and exploited.

THE IMPORTANCE OF METALS

Thousands of years ago, early man first began to make use of minerals. This began with the use of tools and weapons made of flint in what is called the Stone Age.

Fig 11.1 (a) Aluminium is the most abundant metal on Earth, but bauxite is the only workable ore. Bauxite is aluminium oxide.

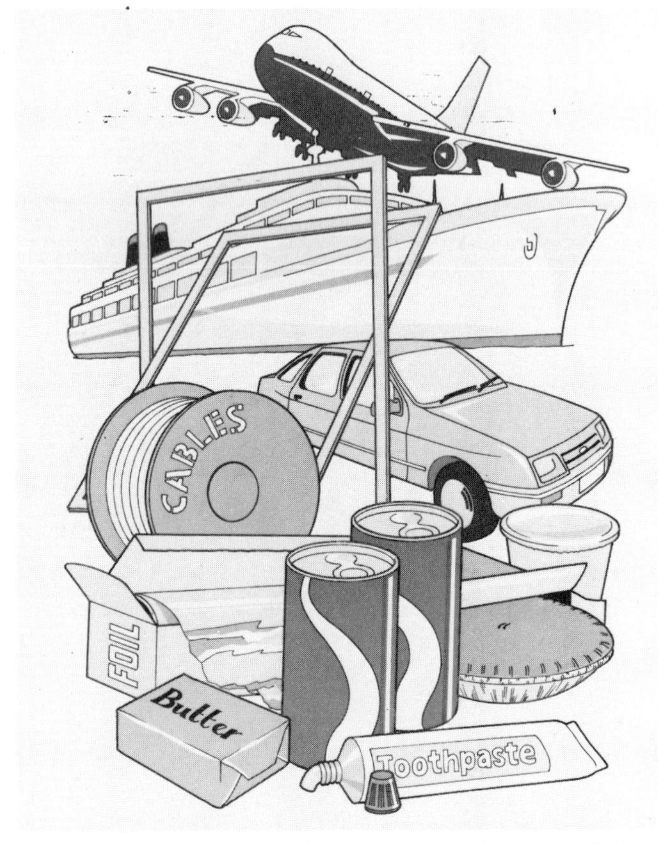

Fig 11.1 (b) Aluminium is a light and strong metal. It can easily be rolled into thin foils and it can be extruded into complicated shapes.

Some of the earliest metal ores were minerals containing copper, eg, malachite. It is still used today in the electrical and chemical industries. It is easier to *smelt* or extract than most metals. Its minerals only need to be heated to 1000°C. It is possible that the first globules of metal were found by accident on the surface of stones next to a roaring camp fire.

Copper gave those who used it a great advantage. They could mould it easily into tools and vessels. This was a quicker process than chipping flint for hours on end. However, one drawback for weapon-making was its softness.

Then came the discovery of tin. Tin is also a soft metal but when added to copper the combined metal, or *alloy*, is much harder than either copper or tin. This alloy of copper and tin is called *bronze*. It was discovered about 5800 years ago somewhere in the Middle East. That time in history is known as the Bronze Age.

The discovery of iron was another important advance. The Iron Age began around 3500 years ago. Iron needs much higher smelting temperatures than tin or copper. Its melting point is about 1500 °C. When mixed with small amounts of carbon, a new alloy *steel*, was found. By 1000 BC steel was being made in India, but it remained rare and little used until the eighteenth century. (Samurai swordsmen perfected the use of steel for weapons in the twelfth century.) A sword made of steel was flexible, yet very hard. Those using steel had a definite advantage. Iron is still the most widely-used of all the metals. Every year the world produces over 305 million tonnes of iron. Twice that tonnage of ore has to be mined. Haematite is the most common ore.

Nowadays we depend on many different metals. *Aluminium*, for instance, is extracted from its ore *bauxite*. Bauxite is a *residual* ore. It forms when aluminium-rich rocks are weathered in a tropical climate. As the rock is softened and breaks down, all the soluble products of weathering are carried away by water. Aluminium oxide is insoluble and is left behind as bauxite. Aluminium is strong, light and it does not rust. This useful metal finds its way into aircraft, ships, cars, window frames and greenhouses. It is also used to make overhead electrical cables, food-cans and wrapping foil for cooking (Fig 11.1b).

Prospecting

Prospecting is the search for new wealth in underground deposits of minerals. At one time the discovery of new deposits of diamonds, gold and ore minerals was largely a matter of luck. It led to the famous Gold Rushes in North America and Australia in the 1800s. Prospectors would spend months of hard work *panning*

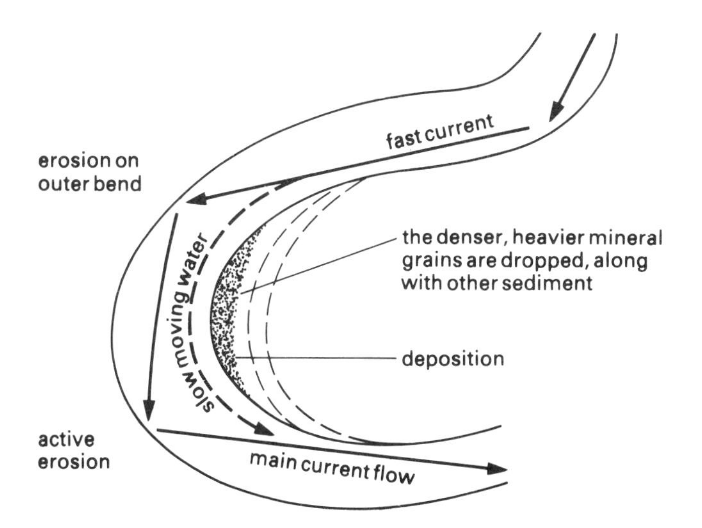

Fig 11.2 A stream placer forming on the inner bend of a river

for gold in the sand of a river bed. Occasionally they would be lucky and find something. This would perhaps lead them to a vein of gold in the rocks further upstream, from which the gold had been eroded.

Placers

When a steam has cut through rocks that contain ore minerals, they are eroded out and become part of the transported sediment.

Any ore or gemstone removed and transported by processes of erosion can form deposits in sands and gravels on beaches and in river channels. Such deposits are known as *placers.* They are at the surface and are usually easy to work.

In a stream, minerals with densities greater than 3 separate from the lighter minerals and settle quickly. They collect in bands and layers, and are often found concentrated on the inner bends of streams. Here the current slows and drops the denser minerals (Fig 11.2).

Beach sand can also contain heavy minerals. Wave action will often sort material. An incoming wave brings mineral material up the beach. The backwash from the wave returns the lighter sand down the sloping shoreline, but does not have the energy to wash down the heavier denser ores. They become concentrated in a zone on the shoreline.

Panning for ores

Panning is a simple technique for separating denser ore mineral grains from the lighter unwanted minerals. You too can become a prospector!

It depends where you are as to what you might find but common ore minerals are:

Ore	*g/cm^3*
galena (lead)	7.5
sphalerite (zincblende)	4.1
cassiterite (tin)	7.0
magnetite	5
haematite	5

Activity 1: Practising the technique of panning

1 Take any sort of flat-bottom pan, 10 cm deep and 20–30 cm in diameter, with sloping sides. The inside of the pan needs to be quite smooth. You can simulate the technique in the following way.
2 Place 10 small lead ball bearings or fishing weights in the pan. Half-fill it with sand and thoroughly mix. Submerge the container in a large container of water and 'pan' the mixture. Can you find all 10 balls in the final two tablespoons of sand?

Alternatively, if you have plenty of galena available, crush up the mineral into small fragments and mix it with the sand in a pan. Then submerge and pan the mixture.

Activity 2: Panning for mineral ores

Submerge the pan below water level on a beach or in a river, and scoop up some sand and gravel. Half-fill the pan then, keeping it submerged, start to rock and rotate it. This action brings the lighter, muddy, material to the surface. Allow it to slop over the edge of the pan. The heavy minerals and pebbles will settle to the bottom. Any unwanted pebbles can be picked out and the process continued.

Eventually the water in the pan will become clear and you are left with the ore minerals.

MODERN PROSPECTING

Prospecting and exploration is nowadays a much more exact science. Rocks are surveyed for hidden ore deposits, using a wide range of sensitive instruments. These can detect variations in density, electrical conductivity, magnetism and radio-activity caused by the minerals underground. One time-saving technique is the use of photographs taken from orbiting space satellites. Such high-detail photographs often show up huge mineral deposits, and greatly speed up the process of mineral exploration.

Density variations

Ore bodies not visible at the surface can often be found by measuring tiny variations in the force of gravity. Denser rocks will give a slighter greater gravitational pull. Structures like salt domes, made of less dense rocks will give a smaller gravitational pull (Fig 11.3). A *gravimeter* is used to measure these tiny variations. A sensitive weighted spring is housed inside a vacuum chamber and kept at a constant temperature and pressure. The reading is recorded by a meter at the end of the eyepiece (Fig 11.4). The instrument must be placed on a level surface for the reading to be accurate.

Seismic prospecting

Seismic prospecting is widely used to map out rock

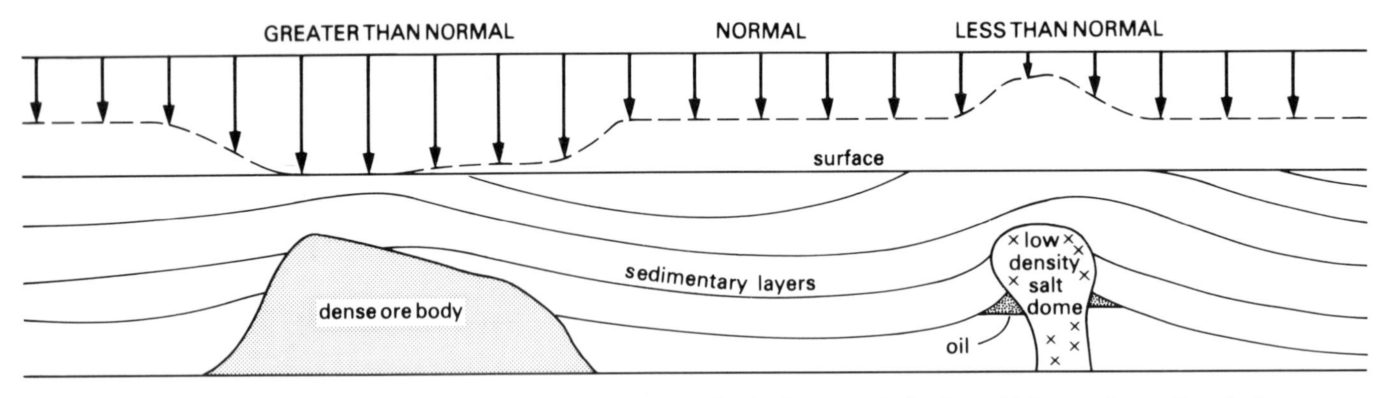

Fig 11.3 A gravimetric survey reveals small variations in gravitational pull. The denser ore body gives a higher gravity reading; the lower density salt dome gives a lower than normal gravity reading.

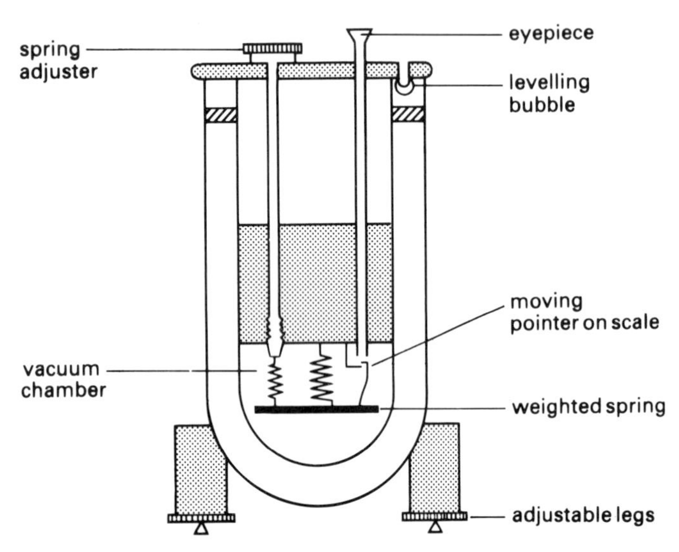

Fig 11.4 A gravimeter

structures below the ground, in the search for oil and mineral deposits. An explosion, or man-made vibration sets off shock waves, which bounce off the rock layers and are recorded. (This method is explained more fully on page 135.)

Magnetic surveys

Magnetic surveys are useful for quickly detecting magnetic iron ores. An aircraft carries an instrument called a *magnetometer*. This gives a continuous recording of the Earth's magnetic field. The presence of magnetic ores will show up as peaks on the recording trace.

Geochemical surveys

Geochemical methods of prospecting may be used to find unusual concentrations of metals in soil or plants. Soil forms by the breakdown of the rocks beneath it. If a soil forms from the rock in an ore body, then it will contain abnormally high amounts of the metal and so will the plants in the metal-rich soil. Chemical analysis of soil and plants can indicate an ore body nearby (Fig 11.5).

Assessing the value of an ore deposit

Whatever the first method of surveying used, the final stage is to take borehole samples. The samples enable the geologist to determine the size of the ore body and its quality. Many ore deposits remain unexploited because of their small size and poor quality. Even good quality ore may be rejected because it is too deeply buried or too far away from roads or railways and so costly to extract. In other words the value of the excavated ore must be *greater* than the cost of extracting it.

EXTRACTING THE MINERALS

When ore deposits have been found, they must be extracted in the most economical way possible. The actual method used depends on the nature of the ore and its depth below ground.

Surface ores: placers

A number of different ores and gemstones occur as surface *placer* deposits, notably gold, tin and diamonds. In a placer deposit, natural water movements have concentrated the ore in a particular locality, making it economic to extract.

Panning is one method of extracting a placer mineral, but larger-scale washing methods are used nowadays. For example, high pressure water hoses are used to extract tin in soft placer deposits in Malaysia. The jets of water wash away the sediment, leaving behind the heavier tin ore. The ore is then refined to extract the metal.

Near-surface ores

If the ore deposit is near the surface, it is possible to extract it by *open-cast mining*. The unwanted layers of rock are removed and the ore is dug out by giant excavators. Open-cast mining is usually carried out on a large scale and the mines are not very attractive features of the landscape. (See page 131 for more detail of this method.)

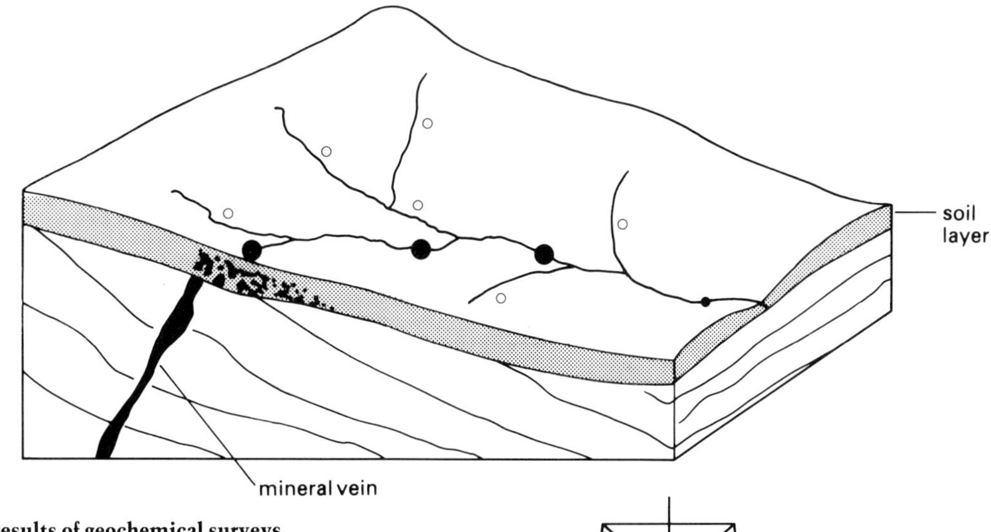

Fig 11.5 A diagram illustrating the results of geochemical surveys

Mining

If a mineral ore is to be extracted from deep below the surface, then shafts and tunnels have to be sunk to reach it. (See coal mining page 131).

Solution mining

Minerals like sulphur, phosphorus and rock-salt can be extracted by drilling boreholes. Hot water and steam are pumped down the outside shaft of the borehole. The mineral dissolves in the hot water and the solution is then sucked up through the central pipe (Fig 11.6). After extraction, the solution is treated so that the water evaporates away, leaving behind the mineral.

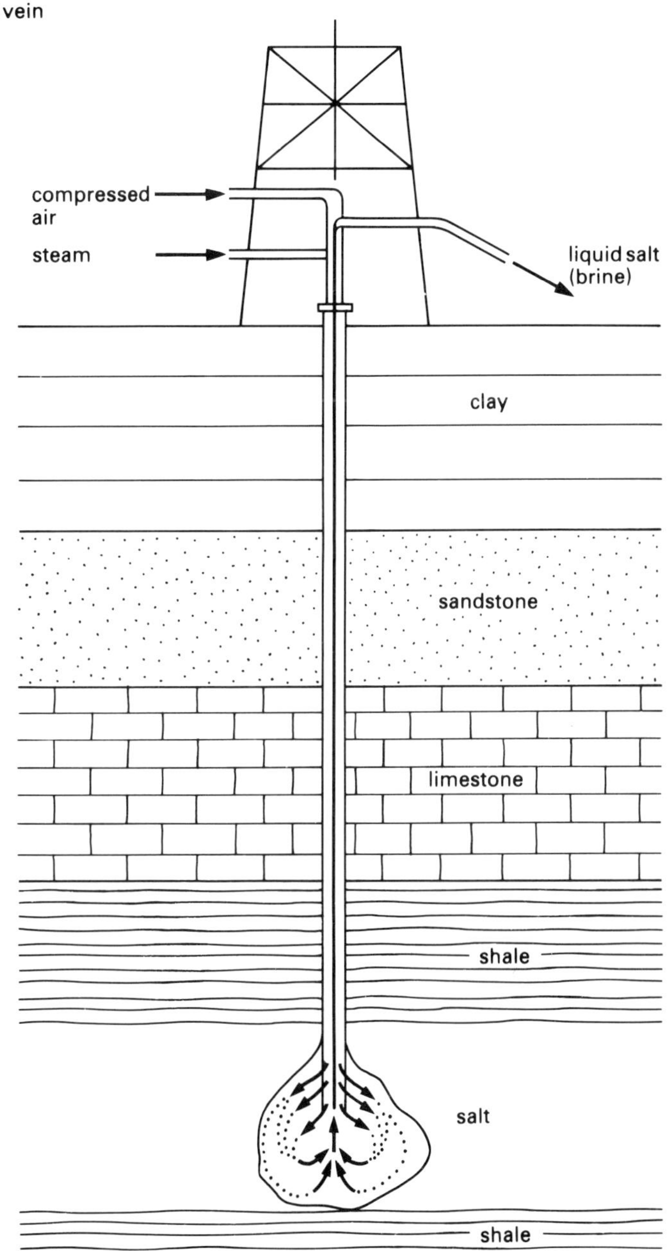

Fig 11.6 A salt well. Steam and hot water pass down the outside shaft. The salt dissolves in the water. The brine (liquid salt) is then forced to the surface.

Exercise 1: 'Prospecting for treasure'

Design a game you could play with your friends. First plan a route outline of numbered squares. Then fill in some of the squares with Rewards and Penalties. Some suggestions are shown below, but you could make-up your own as well.

You find a gold placer. Move to 14.
Your panning equipment is stolen. LOSE A TURN.
Your mule dies. LOSE A TURN.
You find some GOLD NUGGETS! Move forward to . . .
Unless you throw a 6 your nuggets turn out to be FOOL'S GOLD! Move back to . . .
You are rich! Buy a gravimeter. Move forward to . . .
You catch malaria. LOSE A TURN.
You may have discovered uranium. Throw a 4 or 6.
You develop RADIATION sickness! LOSE TWO TURNS.
You are left £10 000!
ODD THROW – you gamble it away – LOSE A TURN
EVEN THROW – move forward to . . .
You discover a Copper Mine. Move forward to . . .
You lose your way in the desert. Move back three spaces.
A throw of 2, 4 or 6 opens up a new DIAMOND MINE! Move forward to . . .

A throw of 2, 4 or 6 means the diamond mine is a success. Have an extra turn!
You open a new TIN MINE! but the price of world tin drops. If you throw 1 or 5 go back to . . .
WAR closes the diamond mine. Go back to . . .

Exercise 2

1 Pair the following phrases together, and write them out as correct sentences.

a) The first tools and weapons	were made of copper.
b) Bronze is an alloy of	iron and carbon.
c) The melting point of copper	is 1500 °C
d) Bronze was discovered	3500 years ago.
e) The earliest metal tools	were made of flint.
f) Iron is an alloy of	copper and tin.
g) The melting point of iron	is 1000 °C.
h) Iron was discovered	5800 years ago.

2 The following passage is a geologist's report about a deposit of aluminium ore. It contains 5 underlined mistakes. Write out the passage, and correct the mistakes made.

Aluminium is melted down from its ore called haematite. This deposit is a resident ore. It has been formed by weathering. Iron-rich rocks have been softened. I expect this happened in a cold climate. Aluminium gets concentrated because it is soluble.

3 Write out and complete these statements. Use the words listed below.

a) When ________ cut through ore bodies, the heavier minerals are removed.

b) Concentrations of ores or gemstones in sands and gravels are called ________.

c) Minerals with ________ greater than 3 tend to settle out much more quickly.

d) The technique for removing ore minerals from loose sediment is called ________.

e) A loose ore mineral found in a stream bed is a valuable clue to the location of a mineral ________ of the ore in the rock further upstream.

vein streams panning placers densities

Exercise 3

1 Outline the main methods of modern prospecting. Explain how they improve on the older traditional methods.

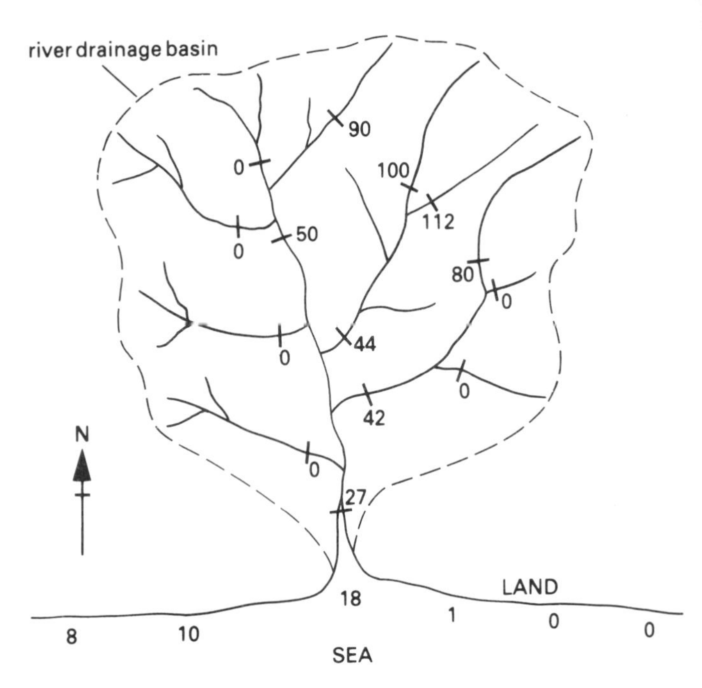

Fig 11.7 Map of the results of a geochemical survey. The figures show the concentrations of copper in parts per million (ppm).

2 Explain fully how:
a) stream placers form;
b) beach placers form.
Include sketches in your answers.

3 a) What is a gravimeter?
b) Look at Fig 11.3 showing variations in gravity. What causes these variations?

4 A survey was carried out to find a copper deposit (Fig 11.7). The survey began along the shoreline. Increasing concentrations of copper were measured in a traverse from west to east. Measurements were then taken in the river basin. There were increasing concentrations of copper along certain streams. The measurements allowed the survey team to identify a mineral vein running across the area.
a) Copy Fig 11.7.
b) Mark in where you think the mineral ore vein is likely to be.
c) Why do the concentration figures get less downstream?
d) What is the main direction of offshore ocean current, from the east or west?

5 Explain, using examples, the main methods of extracting surface or near-surface ores.

12 SEAS AND OCEANS

The oceans cover 71% of the Earth's surface and 80% of the oceans are over three kilometres deep. This is a vast area of our planet. The oceans have great future economic importance. As land-based mineral ores run out, it will be necessary to look for them in the sea.

THE OCEAN FLOOR

Erosion of the land produces sediment. This settles out on the ocean floor. Generally the coarser material is *deposited* (dropped) near to the shore. Only finer muds find their way far out to sea. There are three main zones of deposition (Fig 12.1).

The shoreline zone

Beach deposits of sand and pebbles are common sediments. These may have been eroded from cliff-lines by wave action. Rivers bring down huge quantities of material which frequently collect in layers at the river mouth. A structure called a *delta* is formed. Waves, tidal movements and ocean currents help to distribute and sort the sediment. Layers of pebbles, sand and mud are deposited.

The shelf-sea zone

Most sediment in the shelf-sea zone is sand. Layers of mud are found near to the estuaries of rivers. There is a great variety of life in the shallow, sunlit water and on the seabed. The soft sand and mud is home for many burrowing animals with shells. Their fossil remains are added to the sediments.

Coral reefs

Coral reefs are only found in clear tropical shelf-seas, eg, the South Pacific and Indian Oceans. The coral is a small colonising animal. It needs clear, well-oxygenated, warm water (25–29°C) to flourish and survive. Thousands of individual coral polyps help to make the reef structure. Corals remove dissolved calcium carbonate out of seawater and deposit it to make the reef structure. Lime-secreting algae also help to bind the reef together. The living coral animal sits on top of the solid calcium carbonate it has deposited. Coral reefs grow best in shallow water; a depth of 90 metres is ideal. This means that many shorelines and islands are fringed by reefs (Fig 12.2). The Great Barrier Reef off the east coast of Australia stretches for thousands of kilometres along the shoreline. It is so large that it is easily visible from outer space.

There are also circular islands made entirely of coral. These are called *atolls*. Here the coral depth is much greater than 90 metres. How did the islands form? One theory is that coral began to form a reef around a volcanic island. Then either the island gradually sank, or the sea level rose. The coral kept growing upwards, enclosing a lagoon. In this way, depths of coral to as much as 1600 metres can be explained.

Fossil reefs are found in many rock sequences. They were formed by coral growth millions of years ago. Coral reefs are important as rock-builders. They eventually form large masses of reef *limestones*.

The deep sea zone

Earthquakes may cause sudden movements of rocks below the sea bed; these can disturb thick layers of loose sediment on the continental shelf. Dense clouds of sediment-laden water then roll down the continental slope. These underwater avalanches are known as *turbidity currents* (Fig 12.1). The sand and mud settles out in layers on the deep sea floor. Over the years this process repeats itself many times. Each

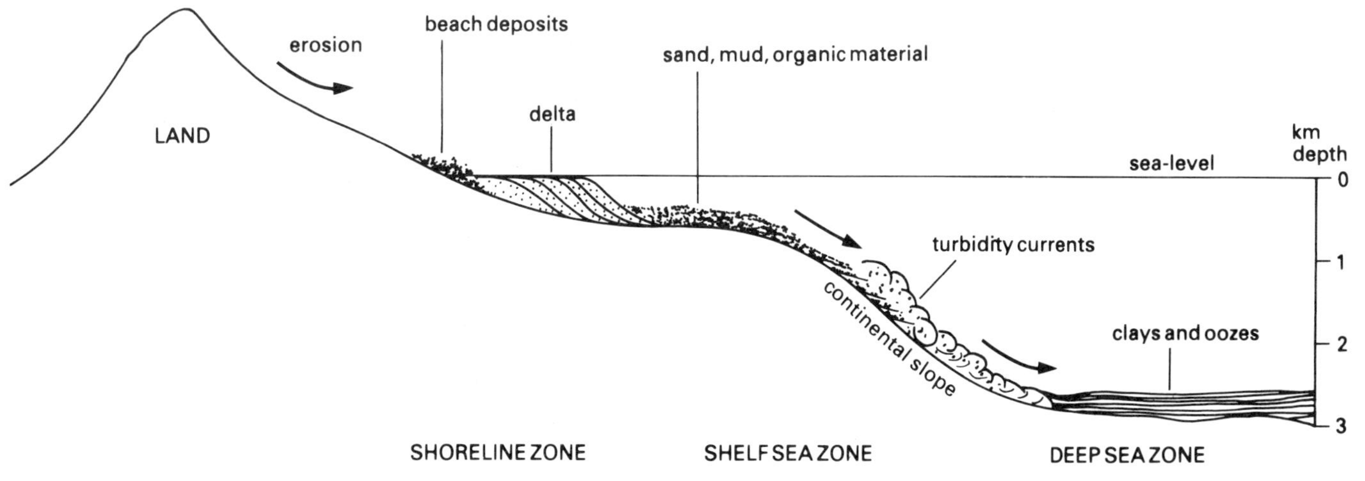

Fig 12.1 Sediments on the ocean floor

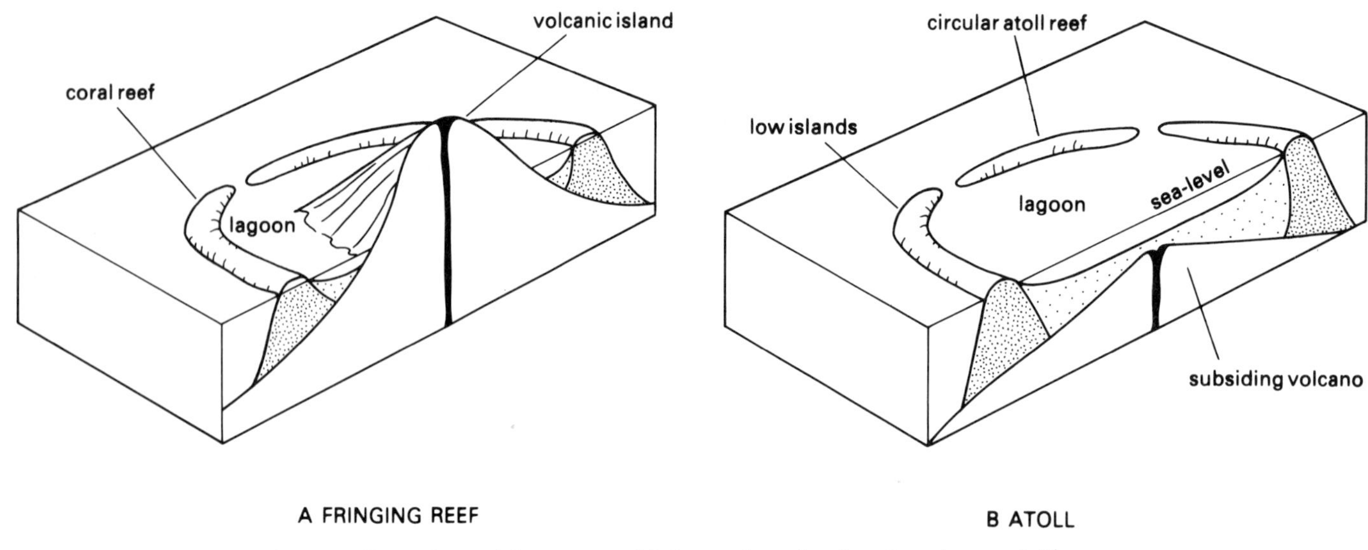

Fig 12.2 Fringing coral reefs and atolls. Reef growth keeps pace with the rise in sea level, as the volcano subsides

turbidity current carries millions of tonnes of suspended material. As the material descends, erosion gullies are carved out of the continental slope. These are known as *submarine canyons.*

Few animals can live at 4000 metres on the deep sea floor. The water is very cold and the sunlight cannot penetrate that far down. Fossils are formed in deep sea sediments. They come from the skeletons of floating or swimming creatures that have dropped to the bottom after death. There is an endless steady 'snowfall' of the shells of tiny floating organisms. Their remains often cover the deep sea floor in a blanket of fine *ooze* (mud). *Calcareous (lime) ooze* comes from the shells of microscopic unicellular (single-celled) animals called *foraminifera.* It forms the most widespread layer on the deep sea floor. *Radiolarian ooze* comes from the remains of a floating unicellular animal. It is made of silica. This ooze is usually found in deeper water.

Red clay is another deep-sea deposit. It is wind-blown dust. Much of it has come from volcanoes or meteoric particles. It is made of iron, manganese, and nickel. It slowly drops to the ocean floor and settles as fine ooze.

Activity 1: Turbidity currents

Fill a deep tank with water. Use a wedge block to represent the continental slope. Put sand and mud into a beaker and mix with water until you have a sloppy liquid. Lower the beaker into the tank and upturn below water level. A turbidity current will then flow down the slope, and settle out.

a) How long does the sediment take to settle?
b) Can you see any layering of sand and mud?

Metallic nodules

Iron, manganese, and nickel are also found in the form of nodules. Brownish-black nodules of manganese and iron oxide are known to exist on the sea bed of all the deep oceans. Each nodule is about the size and shape of a potato (Fig 12.3). Besides manganese they contain copper, cobalt and nickel. The process by which they form is not fully understood. They seem to grow by the addition of new layers of metallic oxide.

These nodules have great economic potential and will undoubtedly become an important mineral resource in the future. It is estimated that tens of millions of square miles are covered with nodules. They have a metallic content sufficient to supply copper, cobalt, nickel and manganese to last the world for at least a thousand years.

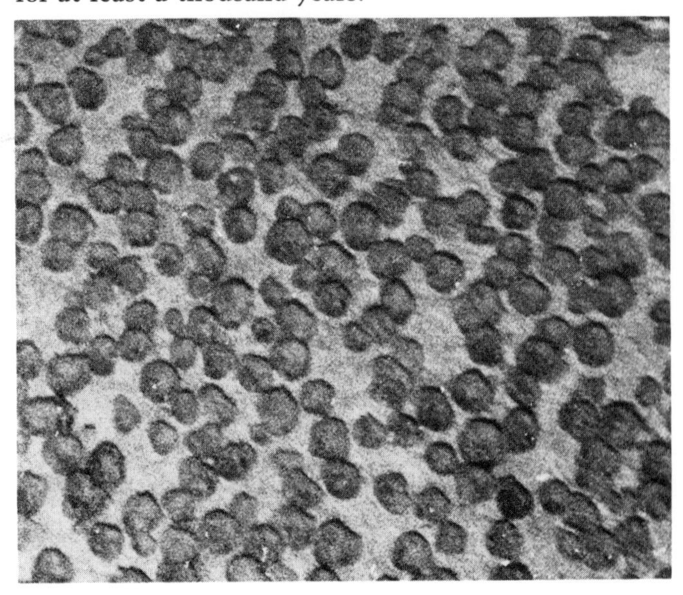

Fig 12.3 Manganese nodules on the deep ocean floor

Submarine 'smoker' vents

New mineral deposits can form in the rocks on the ocean floor. The rocks of the sea floor are pulled apart by plate movements. Molten lava often pours out through these cracks (see page 45). Hot vents and cracks in the sea floor also produce huge amounts of volcanic gases. These are known as black *smokers* (Fig 12.4). The sulphur from these vents reacts with seawater. Bacteria can feed on the sulphides that are produced and other animals live on the bacteria. Giant clams and tube worms have been found. They live in total darkness on the sea floor and their only source of food and warmth is the smoker vents. It is possible that huge amounts of seawater could circulate through faults in rocks beneath the sea bed. Hot water rises by convection through the cracks and faults above the magma chamber. Dissolved minerals from the heated seawater are deposited as they pass upwards through the rocks. Copper, zinc, silver, cadium and cobalt have all been found at smoker vent sites.

THE SALTY SEA

Many dissolved minerals are added to seawater by rivers. These come from the breakdown of rocks inland. Many of these dissolved minerals are extracted by living sea creatures who use them to make their shells and other hard parts. They may also be absorbed by freshly-settling sediment. Sodium chloride (common salt) is not used by sea-living organisms. It therefore forms the dominant mineral in seawater: 1 kg of seawater contains about 35 g of dissolved minerals, 30 g of this is sodium chloride.

Table 1 Dissolved substances in seawater

Substance	%
Sodium chloride	85%
Sulphates of magnesium, potassium, calcium	13%
Many trace elements	2%

Nearly all the elements are found dissolved in seawater. Most of them are present in only tiny concentrations. For example, there is one hundred times more gold in seawater than has ever been found on land. The Germans once tried to extract this gold but they failed because it is in such low concentrations (0.000004 parts per million). It was too expensive to produce.

Formation of saline rocks (evaporites)

In some desert regions, there are vast deposits of salts in dried-up lakes and sea inlets. These saline rocks, called *evaporites*, form when the salts become concentrated enough to be deposited (Fig 12.5). A very hot dry climate is necessary to produce a high rate of evaporation. Conditions like this exist today near the Red Sea, and the Dead Sea in Israel.

In the Permian and Triassic periods (250 million years ago) evaporite salts were deposited in Britain (Fig 15.7). At that time Britain had a hot dry climate. A gulf inlet of the sea became cut off and dried up in the desert conditions. Up to 25 metres of rock salt, gypsum and other evaporites were deposited. These salt deposits are now mined in Cheshire. It would seem that seawater must have refilled the basin several times in order to produce such thicknesses of salts.

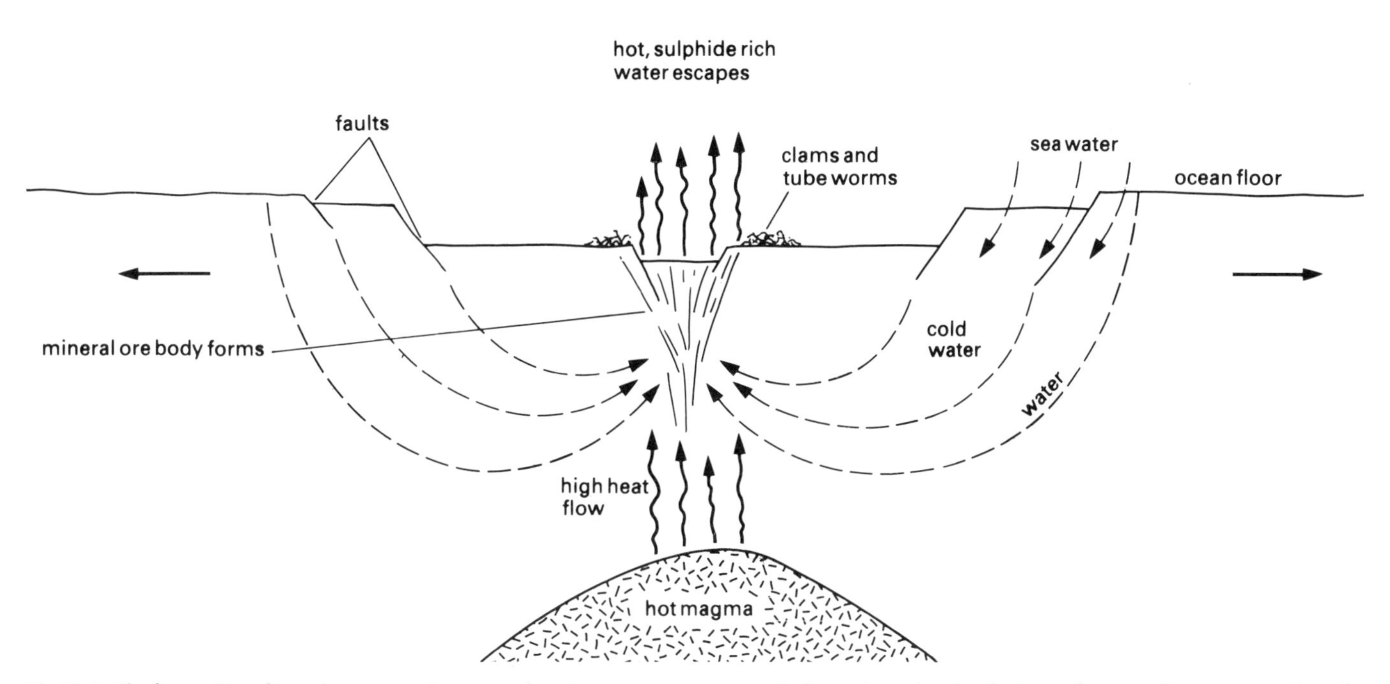

Fig 12.4 The formation of 'smoker' vents above a magma heat source. Seawater is drawn into the circulation and convection system. Minerals come out of solution and they are deposited to form a new ore body.

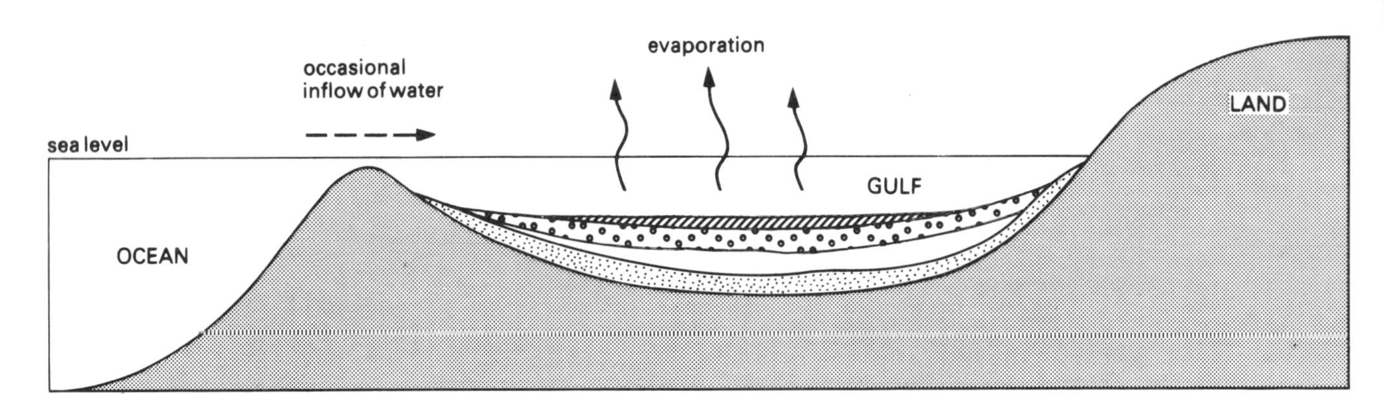

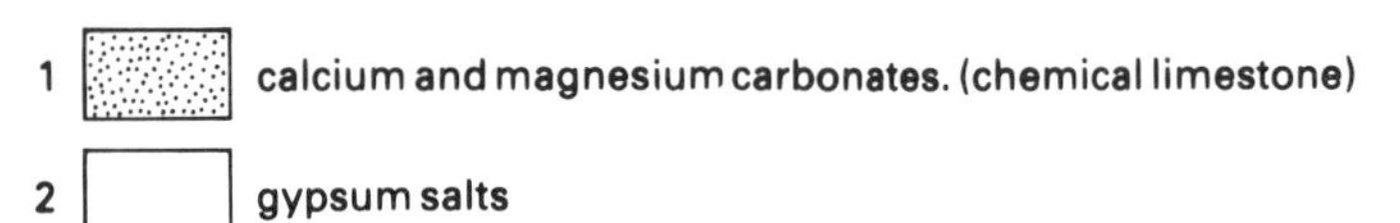

The salts are deposited in the order shown in the key (1–4). This happens as the gulf water evaporates and becomes concentrated. Crystals of salts form and settle in layers on the gulf floor.

Fig 12.5 The formation of evaporites

Salts like gypsum have many uses, (Chapter 7, Table 5).

About a third of the world's salt is produced by evaporating seawater. Borax, a source of boron, is also produced in this way. About two-thirds of the world's supply of magnesium and bromine is also produced from seawater. All these minerals have many economic uses. Borax, for example, is added to glass, weedkiller and fertilizers, as well as being used to make bleaches, soaps and detergents.

Salinity

Some oceans are more *saline* (salty) than others. If there is a lot of fresh water entering an ocean inlet then the salinity will be low, eg, the Baltic Sea. An enclosed ocean inlet in the tropics, eg, the Red Sea, has a very high salinity.

It is possible to produce fresh water from seawater by a process of *desalination*. The dissolved salts are separated out by electrical and chemical means. Desalination plants are expensive to run but they are economically worthwhile in desert countries, eg, Saudi Arabia.

Activity 2: Evaporating water

Try to find out what kinds of solids are left after seawater has evaporated. First, boil some seawater in an evaporating dish. Do the same thing with river water and distilled water for comparison.

1 Which kind of water leaves the most deposit?
2 What colour is the deposit? Does the deposit taste of anything?

Activity 3: Is the seawater deposit mainly sodium chloride?

If the seawater residue is placed in a flame, and the flame turns yellow, it means sodium is present.

1 Clean a flame test-wire by dipping it in hydrochloric acid. Then dip it into the deposit.
2 Hold the test-wire in the flame. Does the flame turn yellow?
3 Repeat this test using a drop of sodium chloride solution. Do you get the same results?
4 Now dissolve the remainder of the deposit in water.
5 Add a drop of silver nitrate solution. If a chloride is present a white solid will settle out from the solution. Does this happen?
6 Repeat this using sodium chloride solution. Do you get the same result?

Conclusion

Sodium chloride must be the main salt in seawater.

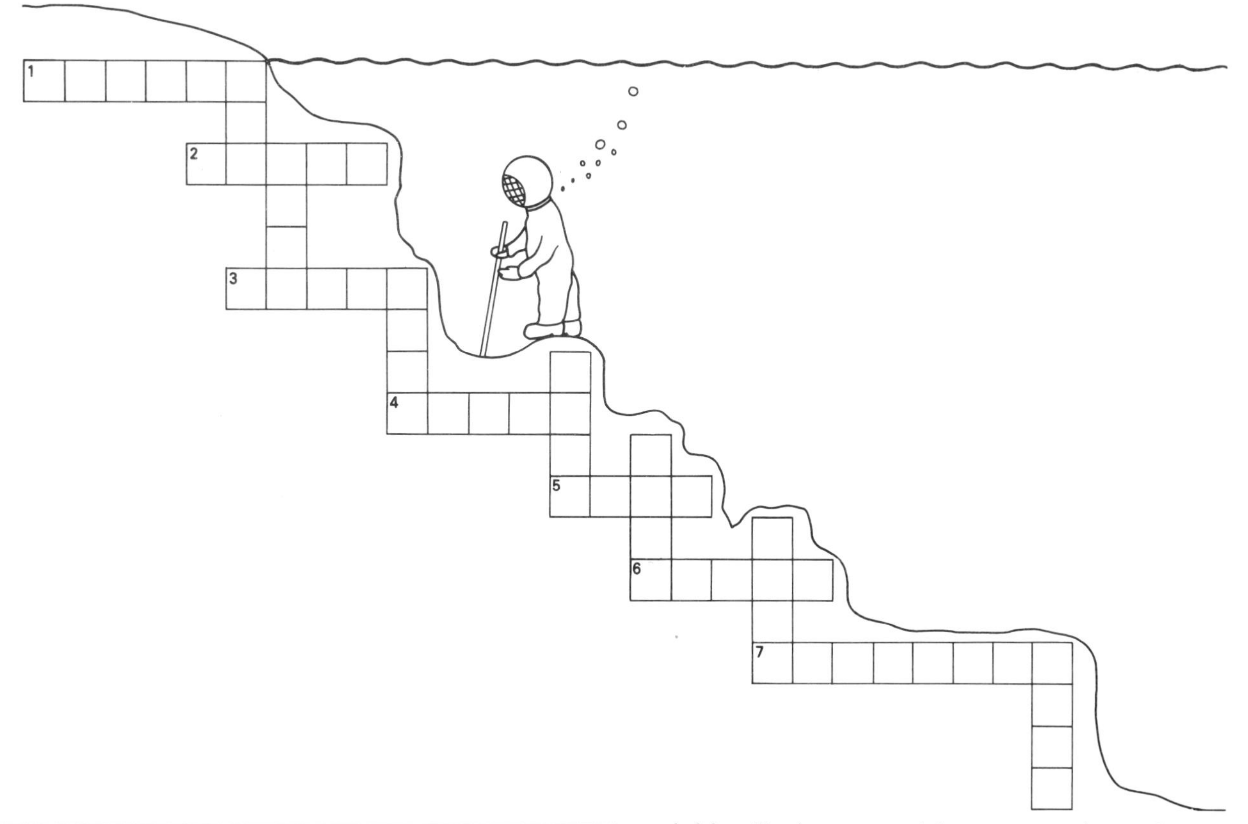

Exercise 1: Diving to the sea bed

The diver is making a survey of the continental slope. Help him to get down to the deep ocean floor! You are only given clues for the horizontal words. Fill in the seven vertical words, selecting from the list below:
reef land vent sea sand salt gold

Clues

1 They cover over 70% of the Earth's surface (6)
2 Cliff lines are eroded by them (5)
3 A coral island (5)
4 It is found at the mouth of a river (5)
5 The opposite of shallow (4)
6 The sea bed, by another name (5)
7 There is no light at this depth (8)

(*Solution on page 143*)

Exercise 2

Write out the following statements correctly. Choose the most accurate word or phrase from those shown in brackets.

a) The action of rivers, ice and wind (erodes/sorts/splits) the land surface.
b) Rivers carry (minerals/sediment/carbonates) to the sea.
c) Sometimes a layered structure called (a cliff/an estuary/a delta) forms.
d) This layered structure is found at the (source/tributary/mouth) of a river.
e) Material that is often carried a long way offshore is (sand/mud/pebbles).
f) A deep sea mud is known as (ooze/clay/sediment).
g) Metallic (cores, nodules, potatoes) are found on the ocean floor.
h) Most life on the sea floor is found in the (shelf sea/continental slope/deep sea).
i) Coral reefs grow best in (lagoons/shallows/sloping shorelines).

Exercise 3

1 Compare the main processes of sediment formation in
   a) shorelines and shelf seas;
   b) the deep sea floor.
2 What are the best conditions for the growth of coral?
3 Look at Fig 12.2 then explain fully how reefs and atolls form.
4 What are metallic nodules? Why might they have future economic importance?
5 Explain fully how volcanic activity and plate movements can lead to the formation of mineral ore bodies in rocks on the ocean floor.
6 'All life depends on the sun as its energy source.'
   a) Explain what is meant by the statement.
   b) Why is the statement no longer absolutely true?
7 Why is it that seawater has a high sodium chloride content, compared to other salts brought in by rivers?
8 What is an evaporite?
9 Why are some oceans more saline (salty) than others?
10 Explain fully the tests that can be done to prove that there is sodium chloride in seawater.
11 Draw a pie (percentage) graph to illustrate the composition of seawater shown in Table 1.

13 LIMESTONE

Limestones are found all over the world. If a sedimentary rock is mainly made of *calcium carbonate* (calcite), then it is a limestone. There are many different varieties. Many of the rocks that are found in England and Wales are limestones (Fig 13.1).

ORGANIC LIMESTONES

Most sea creatures such as corals and shellfish take calcium carbonate from seawater and use it to make their skeletons. When they die, their remains settle on the sea bed. Eventually the shelly fragments compact together. A solid *shelly limestone* is formed (Fig 13.2). Coral reefs also form *reef limestones*. Fossil reefs are found in the Silurian rocks at Wenlock in Shropshire.

Chalk

Chalk is another variety of limestone. It forms the Chilterns and Downs uplands in England. Chalk is very pure calcium carbonate (Fig 13.3). Using an electron microscope it is possible to see what chalk dust looks like. Chalk is mostly made of tiny ribbed plates of calcite known as *coccoliths* (Fig 13.4). Each one is only 0.003 mm across. They form the skeleton of a

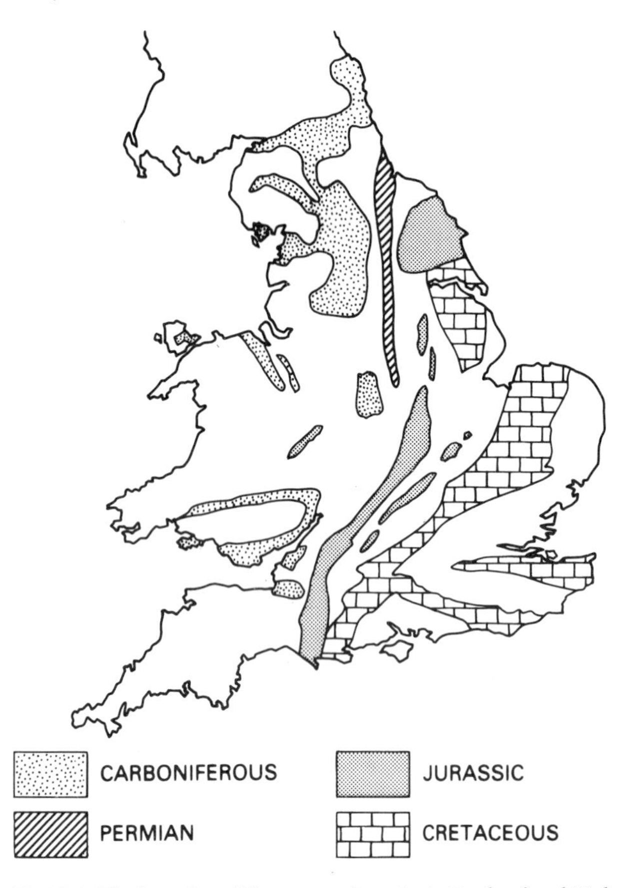

Fig 13.1 The location of limestone deposits in England and Wales

Fig 13.2 Shelly Wenlock limestone. Fragments of trilobites, crinoids, corals and brachiopods can be seen.

Fig 13.3 Chalk cliffs, Burling Gap

Fig 13.4 Coccosphere showing coccoliths

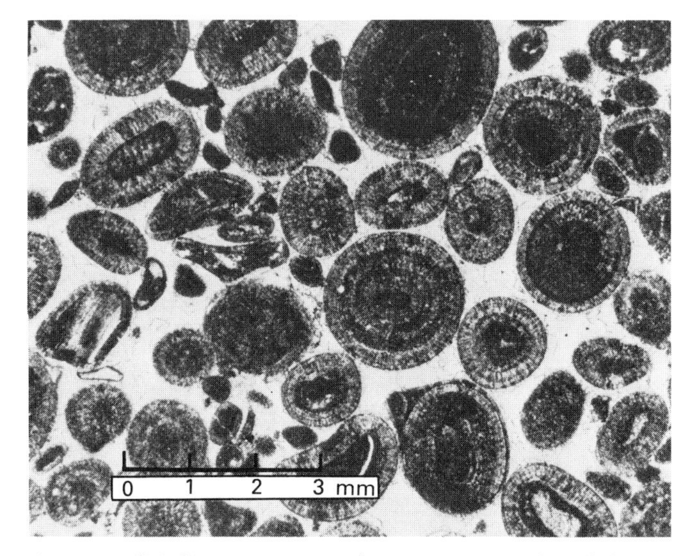

Fig 13.5 Oolitic limestone seen under a microscope. The ooliths are held together by small crystals of calcite.

Fig 13.6 The Great Scar limestone (Carboniferous) above Malham Cove, North Yorkshire

tiny, spherical, floating plant called a *coccosphere*. These plants are still found today in the surface waters of the oceans. There are more than 1 million coccospheres in just 1 litre of sea water! Their coccolith remains collect as a fine mud on the sea bottom. Eventually a chalk rock forms.

CHEMICAL LIMESTONES

It is possible for limestones to form directly from seawater. Living organisms are not involved.

Evaporite limestone

As seawater evaporates, the water gets more and more concentrated (see Chapter 12). The calcium and magnesium carbonates settle out as crystals. The Magnesian limestones of Durham are evaporite limestones. They formed in the sea basin of the Permian sea.

Oolitic limestone

Many limestones are made up of small spheres. These are 1–2 mm in diameter and look like fish-roe (Fig 13.5). They are known as *ooliths* (Greek for 'egg'). Each oolith is made of a number of layered coatings of calcium carbonate; there may be a sand grain or tiny piece of shell at the centre of each one. Ooliths form in the shallow, sloping shorelines of tropical seas. Tidal movements roll the ooliths backwards and forwards in the warm water. Finally, the ooliths collect in layers to form *oolitic limestone*.

Ooliths are forming today in the shallow waters of the Bahamas and the Florida coast. Conditions were very similar in Britain in the Jurassic period (180 million years ago) and most of the Cotswold Hills are made of oolitic limestones of the Jurassic system.

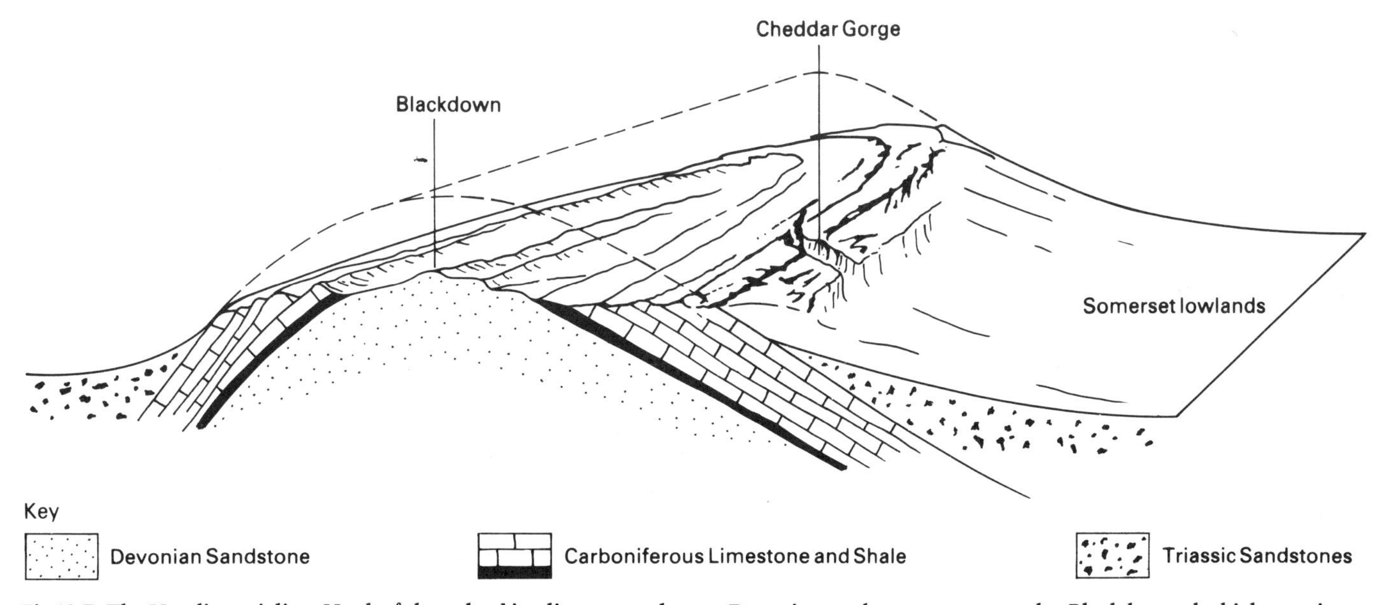

Fig 13.7 The Mendip anticline. Much of the upland is a limestone plateau. Devonian sandstones are exposed at Blackdown, the highest point. The dotted lines show what the anticline was like before erosion.

Some limestones of Carboniferous and Silurian age also contain ooliths.

Limestone scenery

Much of the limestone that can be seen in Britain is exposed as hills and ridges (escarpments). Limestone uplands help to form some of our most striking scenery, for example, the Pennines, the Cotswolds, the Mendips and the Chalk Downs (Fig 13.6).

Carboniferous limestone

Some of the most impressive scenery is formed by *carboniferous limestone*. This particular kind of limestone was formed by sea creatures, 350 million years ago. Much of Britain was then a shallow sea and the conditions were ideal for the growth of corals, crinoids, and many other shelled creatures. Their fossil remains collected to form thick beds of limestone. In time, these beds were folded and pushed up above sea level. They now form the limestone uplands of the Pennines in northern England, and the Mendips in the south (Fig 13.7).

The effects of weathering

As rainwater falls through the air, it absorbs carbon dioxide. This turns rain into weak carbonic acid. The rainwater attacks the calcium carbonate in limestone and changes it into soluble *calcium hydrogencarbonate*. This dissolves easily in water, and so the rocks are slowly eaten away.

$CaCO_3 + CO_2 + H_2O \rightarrow Ca(HCO_3)_2$
Calcium + Carbonic acid → Calcium
carbonate hydrogencarbonate

Limestone is a well-jointed rock which usually only has a thin soil layer, and supports few trees. Rainwater quickly soaks into the rock along the joint planes and bedding planes. One of the most striking features of limestone scenery is the virtual absence of any surface drainage. The joint planes become widened and deepened to form *grikes* with intervening blocks called *clints*. In places, the bare rock is exposed as *limestone pavement* (Fig 13.6).

Surface streams often disappear into vertical holes in the surface, known as *swallow holes* (Fig 13.8). These are formed by the widening of vertical joints. Underground streams flow along joints and bedding planes. The water continues to dissolve the limestone and hollows out caves and caverns. Some of these caves are immense. The Carlsbad Cavern in New Mexico, USA, is 1200 metres long and 90 metres high. Water carrying dissolved limestone drips from the roofs of caves. The limestone is redeposited as *stalactites* (hanging down) and *stalagmites* (growing upwards) (Fig 13.9). Sometimes the roof of a long cavern cut by an underground river collapses and this results in the formation of a *dry-valley* at the surface (Fig 13.10).

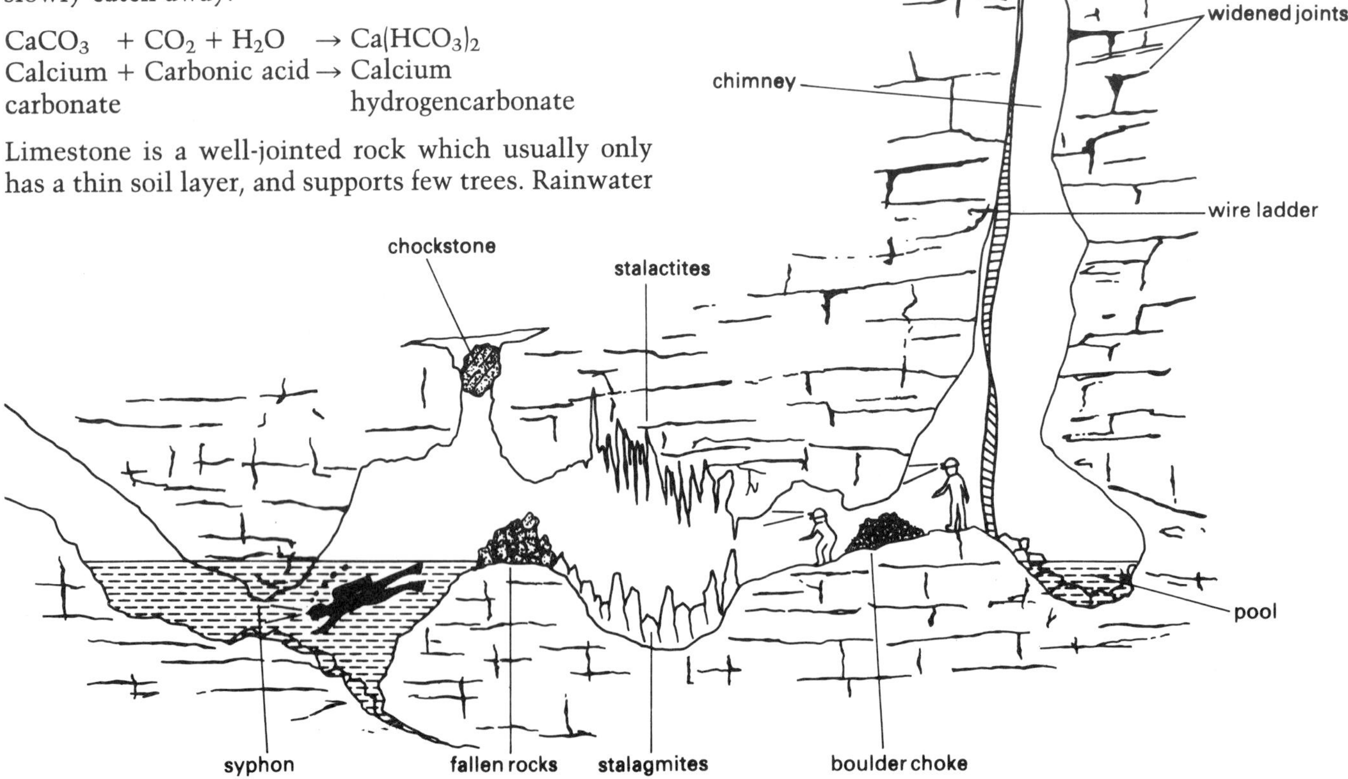

Fig 13.8 Exploring limestone underground

Fig 13.9 Stalactites grow down from the cave roof. This is a very slow process.

Fig 13.10 How Ghyll – a dry valley in limestone, North Yorkshire.

Cheddar Gorge

Cheddar Gorge on the Mendips, is another spectacular example of limestone scenery. The gorge is a dry, narrow valley. It cuts into the southern side of the Mendips. The sides of the gorge form towering vertical cliffs of limestone over 100 m high (Fig 13.11).

The gorge was cut by running water both above and below ground. It seems likely that parts of the gorge formed by collapse of the roofs of former caverns. There are a number of caves in the side of the gorge.

Fig 13.11 Cheddar Gorge, Somerset

Some of these are opened to the public. It is possible to see stalactites and stalagmites, as well as the underground stream which now runs well below the bottom of the gorge. Occasionally the water level rises. The water floods into the caves, and into the gorge itself.

Cheddar Gorge is well worth a visit. (See list of suppliers, page 141.)

Caving (speleology)

Exploring caves is a fast-growing sport. The caver can discover enormous caverns, underground lakes, and raging rivers. Even today, many of the world's cave systems are largely unexplored. Caving requires many skills and techniques. You have to learn the best way to climb up or down vertical drops, using ropes and wire ladders. You have to know how to crawl through narrow tunnels, and avoid falling boulders. Cavers also dive through underwater lakes (syphons) to explore new cave systems beyond (Fig 13.8).

THE USES OF LIMESTONE

Limestone has long been quarried for use as a building stone. The pyramids of Egypt were built using limestone. Many famous buildings in London were built using Portland limestone from Dorset, including St Paul's Cathedral, the front of Buckingham Palace and the government offices in Whitehall. Concrete multistorey office blocks in city centres may not look as if they are built of limestone; but do you know what concrete is made of? It comes directly from broken-up limestone.

How limestone is quarried

Limestone is blasted off the rock face, using explosives. A huge machine called a *faceloader* scoops up the loosened rock and drops it into a dump-truck. The load is then taken to the crusher. Here the rock is broken

Table 1: Some of the many uses of limestone

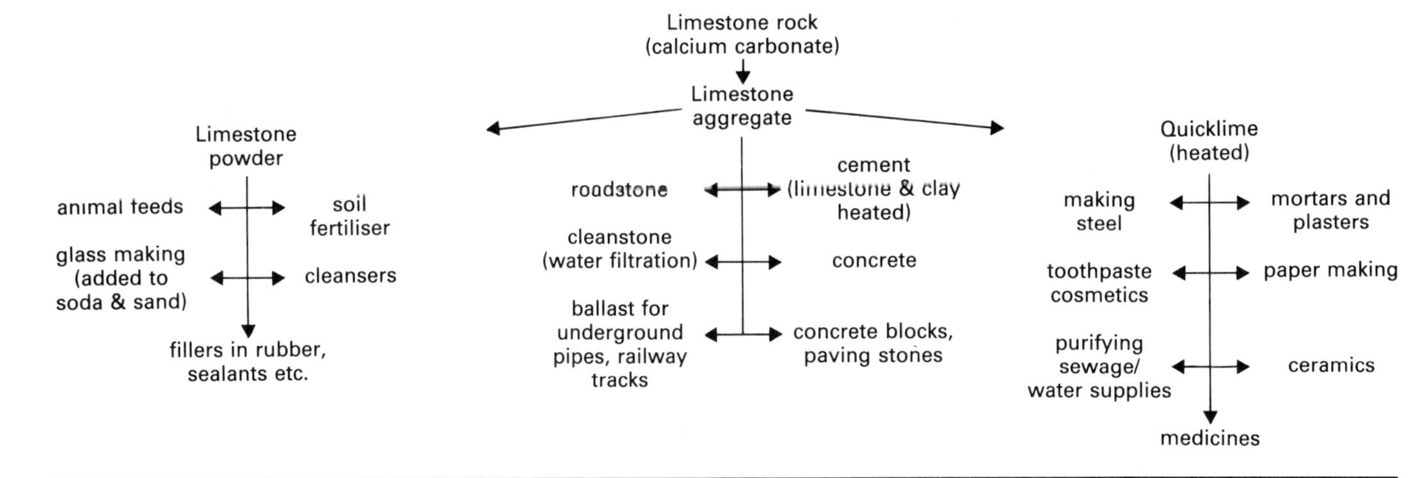

into small pieces. The broken stone is then graded into various sizes from 50 mm down to fine gravel and powder.

What happens to the limestone

The broken stone is known as *aggregate* or *chippings*. Almost half of the aggregate is coated with tar (bitumen). It is then used to tarmac road surfaces. Most of the remaining aggregate is used to make concrete and concrete blocks (Fig 13.12).

Limestone powder is added to animal feeds. It helps to form good strong bones, and cows produce better milk. It is also added to pet food. Your dog could be eating limestone every day!

Fine limestone dust has an amazing number of uses. It is found in foam carpets, floor and ceramic tiles, sealants and mastic fillers. When heated to 1100 °C limestone changes to *quicklime (calcium oxide)*. Quicklime and its by-products have many different industrial uses (see Table 1).

Fig 13.12 Flood barriers. Flexible rafts made of concrete blocks linked by nylon rope. They are placed on sea and river embankments.

Transporting limestone

Limestone is a low-value product. It has to be transported cheaply in large quantities if the quarry is to make a profit. At the large Foster Yeoman quarry, near Shepton Mallet in the Mendips, large lorries are used to supply the local area. The quarry also has a rail link. Huge super-trains transport the limestone to processing centres all over the south-east of England. Each train is over half a mile long and carries over 4000 tonnes of limestone. In this way, the quarry can supply a much larger market area than would be possible using only road links.

Activity 1: Making concrete blocks

1 Mix together: chippings, sand and cement. Work on a ratio of 3:2:1 of each ingredient.
2 Mix with a little water. Be careful not to make the mixture too sloppy.
3 Pour the mixture into a wooden box or frame resting on a sheet of thick polythene. A wooden seed-try with the bottom slats removed is ideal.
4 Smooth off the surface of the concrete and leave to set.
5 Remove the frame and the polythene. You have made a concrete block.

Questions

a) What happens to the strength of concrete if you add
  (i) more chippings;
  (ii) fewer chippings;
  (iii) hardly any chippings?
b) Look closely at the solid concrete block. What kind of natural rock does it remind you of?

Activity 2: Testing for calcium carbonate

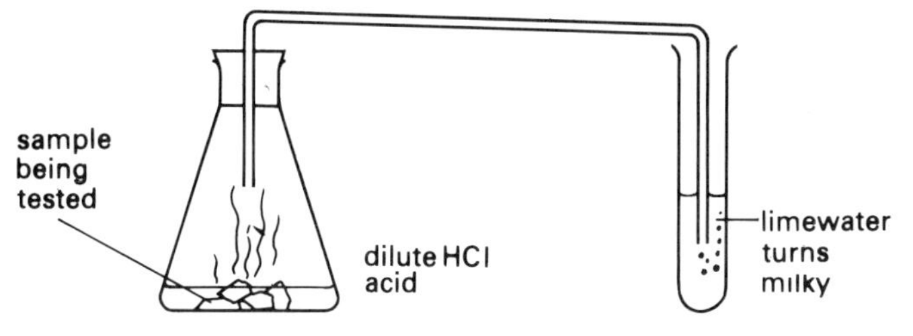

Fig 13.13 Testing for calcium carbonate

Use the equipment shown (Fig 13.13) to test different materials for calcium carbonate. Anything that is calcium carbonate will effervesce (fizz) when placed in dilute hydrochloric acid because carbon dioxide gas is produced. The gas bubbles up through the limewater.

Test some of the following: chalk, brick, oolitic limestone, coral, calcite, eggshell, ceramic tile, sea shell, kettle fur.

Questions

a) Write down the name of any material which does not react to the acid test.

b) The limewater turns milky if there is a reaction. What gas is being produced?

Exercise 1: Let's go caving!

1 Complete this puzzle using the words provided. (*Solution on page 144*)

4 letters	*5 letters*	*6 letters*	*7 letters*	*8 letters*	*9 letters*	*10 letters*
pool	choke	**syphon**	chimney	boulders	limestone	**chockstone**
rock	diver	ladder	surface			**stalagmite**
hole	shaft	helmet				**stalactite**
lamp	water	cavern				
rope		search				
fall						
cave						

2 Explain the meaning of the four **bold** words.

Exercise 2

1 What is limestone made of?
2 Name three kinds of organic limestone.
3 Name two kinds of chemical limestone.
4 Study Fig 13.2.
 a) How many fragments of trilobites can you see?
 b) Make a sketch of one of them. What part of a trilobite is it – the head, thorax or tail?
 c) How many fragments of corals can you see?
 d) Make a sketch of one brachiopod shell.
5 In what kind of limestone are coccoliths found?
6 Make a sketch of one coccolith from Fig 13.4.
7 Using the scale on Fig 13.4, work out the size of a coccosphere.
8 a) What kind of living organism is a coccosphere?
 b) How do their remains form a rock?
9 If ooliths are found in a limestone, what does this tell you about:
 a) the climate;
 b) the place where the rock formed?
10 What were conditions like in Britain in the Carboniferous period?
11 What is Carboniferous limestone made of?
12 Study 13.7. Explain how earth movements and erosion have helped to create the Mendip Hills.
13 What are stalactites and stalagmites? Explain their formation.
14 What is Cheddar Gorge, and how did it form?
15 What is meant by the term 'aggregate'?

Exercise 3

1 Explain fully the formation of:
 a) evaporite limestone;
 b) oolitic limestone.
2 What are the main chemical changes that happen when limestone is exposed to rainwater?
3 Make a labelled sketch of Fig 13.6 to show the following features: thin soil; grass cover; clints; grikes; pavement; bedding planes; dry stone walling; trees and bushes rooted into vertical joint planes.
4 What are the main surface and sub-surface features that develop as a result of the flow of underground streams and rivers?
5 Outline the main stages in the quarrying and production of limestone.
6 Write about the many uses of limestone.
7 Why is cheap and efficient transport so important for limestone products?

14 WATER MOVEMENT

The total amount of water on the Earth is an amazing 1360 million cubic kilometres, and 97% of this water is in the oceans!

THE WATER CYCLE

The hydrological (water) cycle is the circulation of water from the oceans to land areas, and back again (Fig 14.1).

The sun's heat causes *evaporation* of the oceans' surface waters. Water vapour rises on air currents and it *condenses* to form clouds. Winds carry the moist air over the land. The moisture may fall as rain or snow (*precipitation*). Some surface water will enter streams and rivers as surface *run off* and so return directly to the sea. Rainwater also soaks into the soil. Plant roots absorb some of this water. The plants return it to the air via their leaves (*transpiration*). About 22% of rainwater soaks through the soil into the rocks beneath, to become *groundwater.*

Groundwater

The amount of groundwater present depends on the kind of rock. *Permeable* rocks allow water to seep through them. *Impermeable* rocks do not, eg, shales and mudstones. Some permeable rocks soak water up easily like a sponge, the water enters the pore spaces

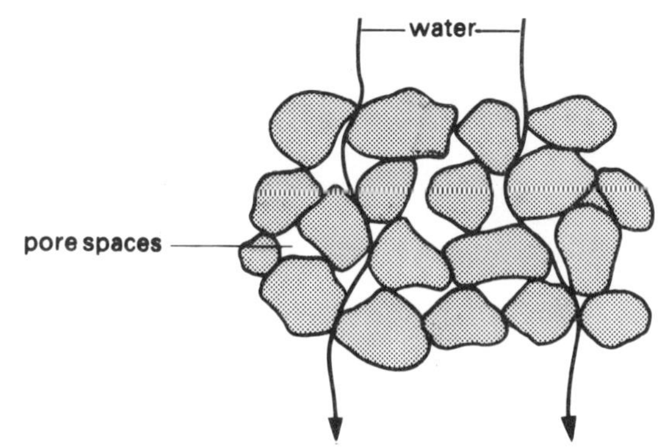

Fig 14.2 Sandstone is porous. It is also permeable (water can pass through the rock).

between the grains. In other words such rocks are *porous*. Most sandstones are highly porous (Fig 14.2). They can hold up to 30% of their volume of water. Rocks like this form good *aquifers* (water bearing rocks). Other permeable rocks, eg, limestone allow water to pass along joints and bedding planes but not through the fabric of the rock. Such rocks are said to be *pervious.*

Water soaks downwards from the surface until it reaches an impermeable rock layer. All the available pore spaces and joints fill up and the rock becomes *saturated*. The top of the saturated zone is called the *water table.* Rocks above the water table may be moist but they will also contain some air-filled pore spaces.

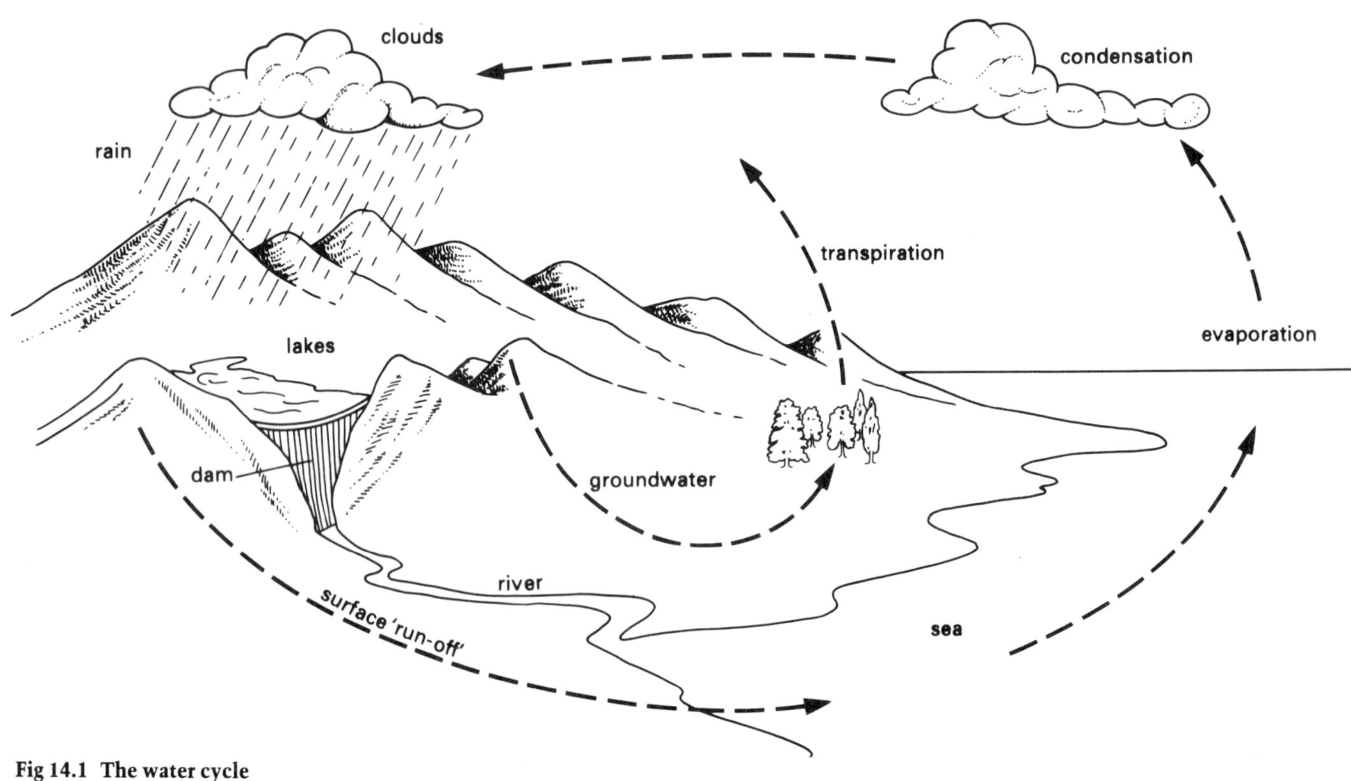

Fig 14.1 The water cycle

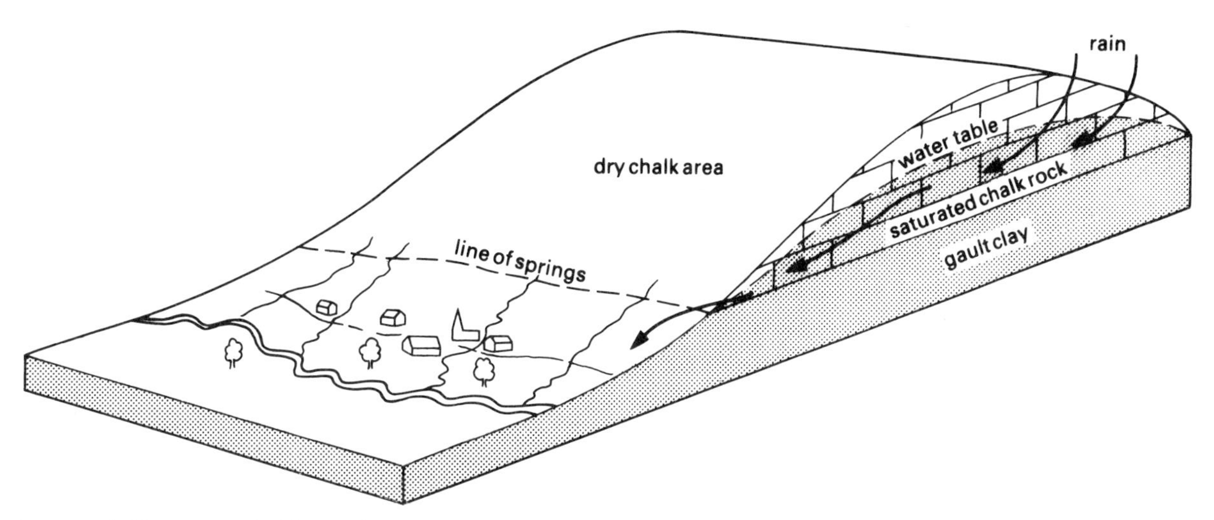

Fig 14.3 Farms and villages as 'springline settlements'. A line of springs is formed where the chalk aquifer overlies the impermeable Gault clay, as in southern England.

Springs

Springs are found where water flows out of the ground again. This happens when the saturated rock is exposed at the surface, with an impermeable rock beneath (Fig 14.3).

Artesian basins

An *artesian basin* is formed when a permeable layer of rock is downfolded in a basin structure. It is sandwiched between two impermeable layers (Fig 14.4). The permeable rock soaks up rainwater at the exposed edges of the basin. The water cannot drain away because of the basin structure. Artesian basins form important *aquifers* beneath the Sahara Desert, and in Western Australia. If a well is sunk the life-giving water gushes out. Water pressure in *artesian wells* is caused by the water table being at a greater height than the top of the well.

THE POWER OF RIVERS

Rivers and streams carry water from the land back to the sea. They start from springs or from melting glaciers up in the hills and mountains. They may also form by surface water that collects in gullies. The downward flow under gravity gives rivers the energy to

Activity 1: How rocks soak up water

It is possible to show how *porous* a rock is, in other words how much water a rock can hold, using this formula:

$$\text{the \% } porosity = \frac{\text{volume of water absorbed}}{\text{volume of rock}} \times 100$$

You will need a block of green *Oasis* (as used by flower arrangers), this represents the 'rock'. It is used instead of a real rock because it soaks up water very quickly.

1 First you need to know the *volume* of the Oasis block. How many cubic centimetres (cm^3) does it contain? To find this, measure – length × breadth × height.
2 Fill a measuring jug with water. Note the volume of water.
3 Allow the water to trickle onto the Oasis. How much water is needed to completely saturate the Oasis block? This is the volume of water absorbed by the block.
4 Now it is possible to work out the porosity of the Oasis, using the above formula.
   a) What is the percentage porosity of Oasis. Is it greater or less than sandstone (30%)?
   b) Will Oasis allow water to pass through it? In other words is it permeable?
5 Submerge samples of granite and sandstone in water. Can you see air bubbles rising? What does this prove?

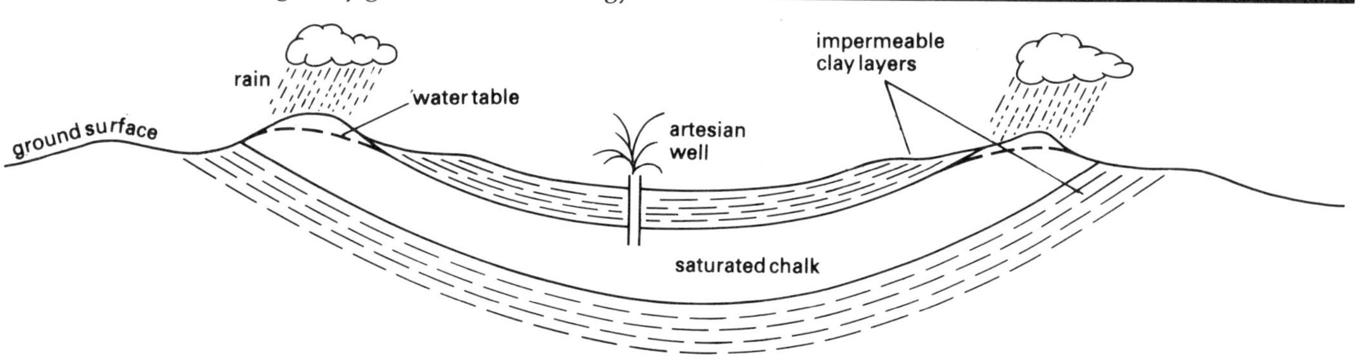

Fig 14.4 An artesian basin

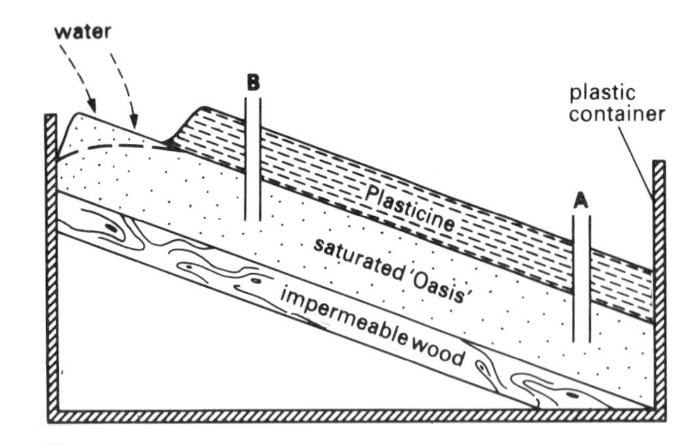

Fig 14.5 Model of an artesian well

Activity 2: A model of an artesian well

1 Cut a slab of Oasis, to act as an aquifer. Lay it on a sloping piece of wood inside a plastic container (Fig 14.5).
2 Put a layer of Plasticine over four-fifths of the Oasis. Ensure a good seal.
3 Pour water onto the exposed Oasis, until it is saturated.
4 Bore a hole through the impermeable plasticine with a pencil at A. What happens?
5 Bore a second hole at B. What happens?

Questions

a) What is the purpose of the sloping wood beneath the Oasis aquifer, and the Plasticine above it?
b) Explain why the water coming out of the two 'wells' behaves differently.

erode steep-sided V-shaped valleys. Loose rocks and pebbles removed by erosion are picked up by the current. They collide with the banks and bed of the river and this causes more downcutting and erosion.

Erosion

Potholes may form on the bed of a stream. They are eroded by the down-cutting, swirling, corkscrew motion of pebbles in eddying currents (Fig 14.6). Eventually, as the separate potholes widen, they join together and the stream bed is lowered. In the upper parts of river valleys there are many waterfalls and rapids. These are caused by hard bands of rock that underlie the river bed and which are eroded at a slower rate than the softer rocks. Lakes occur where the river flows into hollows. Eventually the hard bands of rock are eroded away, and the lakes are filled up with sediment. In this way, the irregularities in the course of a river are removed.

Transport

Rivers transport boulders, pebbles, sand and mud, as well as dissolved material. The material carried by a

Fig 14.6 Potholes eroded in the hard rock of the river bed

Fig 14.7 Deposited load on the bed of the River Roeburn at Wray, Lancashire

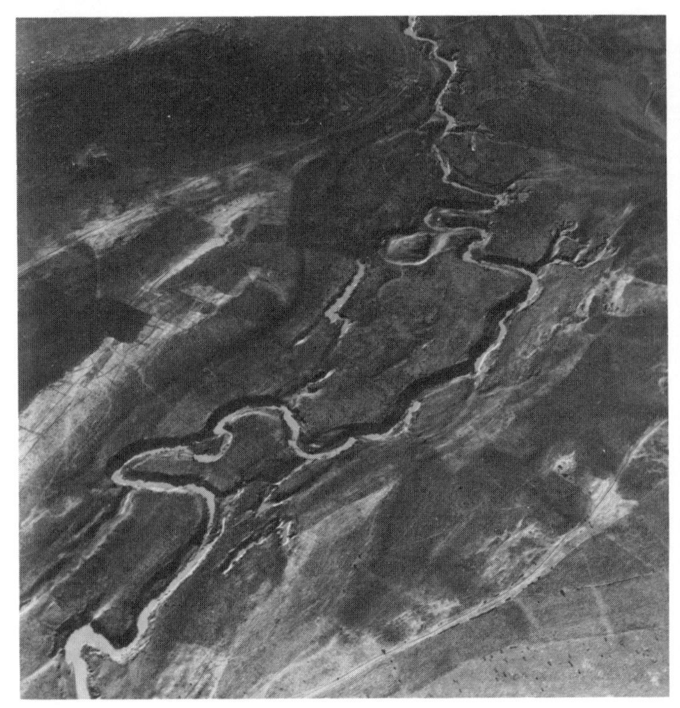

Fig 14.8 The Jordan Rift Valley. Notice the meanders and ox-bow lakes.

river is called its *load*. Rivers transport load by four main methods:

1 *Solution* chemicals that are carried dissolved in the water.
2 *Suspension* silt, mud and fine sand that are carried along without touching the river bed.
3 *Saltation* larger particles that are carried along in suspension, but which sink because of their weight. Then they rebound off the river bed and rise into the faster current again.
4 *Traction* pebbles and boulders that either roll or slide along the river bed.

Deposition

In the lower parts of river valleys the river flows down a shallower slope. The valleys are wider and the river *meanders* (bends) its course from side to side. There is still a good deal of erosion especially on the outer bends of meanders; on the inner bends the current is slower and the load is deposited (Fig 14.7). *Ox-bow lakes* may form if the river cuts off its own meander (Fig 14.8). Near to the river mouth the river may flow over a wide flat area called the *flood-plain*. Here much of the load may be deposited, especially along the river banks where the current is slower. These deposits can build up to form raised banks called *levées*. This process can mean that the river is actually flowing at a higher level than the surrounding flood plain.

The general name given to river sediment is *alluvium*. At the river mouth the sediment may be carried away by ocean waves and currents. If the tides and currents are weak, the river drops alluvium. A layered mound is built out into the water as a delta, eg, the Mississippi delta in the Gulf of Mexico.

Rivers are important agents of erosion. The energy of running water is capable of eroding, transporting and depositing huge quantities of material. The tremendous power that streams and rivers have is shown by the following example:

The Wray Flood, North Lancashire (Fig 14.9)
At 6.30 pm on 8 August 1967, the village of Wray in North Lancashire was badly flooded. Bridges were swept away as millions of gallons of water poured into the streets. A small stream suddenly became a raging torrent.

The flooding of Wray began with a freak cloudburst. About 50 mm of rain fell in the space of an hour on the hillsides above Wray. A wall of water rushed down the River Roeburn, followed by a second surge, a few minutes later. The water brought with it a mass of stones and boulders. Trees were uprooted from the banks. Many were carried downstream, still in an upright position. Several bridges were destroyed. Their stonework was added to the tumbling mass of rubble and boulders hurtling downstream. Three houses were washed away completely and many others were damaged. Cars were also swept away. They were found later several kilometres downstream. Amazingly, no one was killed.

Fig 14.9 The destruction of Wray village by flooding

Fig 14.7 shows the River Roeburn at Wray during a summer drought. Can you imagine this stream tearing away roads, bridges and houses in August 1967?

Ruth Whittam was literally washed out of her house. This is how she described what happened:
'Water rushed into the living room and kitchen. I sat in the sink, but the water level rose to sink level. I climbed onto the cooker, but it started to float away. Then the fridge trapped me against a door. I smashed open the window with a pan, and then I floated out with the water. The next thing I remember I was on the door step. I had collapsed.'

Bill Brown, a farmer said:
'We were dipping sheep when we heard the river careering down at us. We just had time to get out of the way. All our farm buildings were lost. Only one cow out of the herd of 40 was left and only 5 sheep out of 200.'

Why did it happen?

Look at Fig 14.10 which shows the position of Wray. The village is built at the point where two river systems come together. Could you have predicted that some day Wray might be flooded?

The area drained by a river system is called the *drainage basin*. This acts as a *catchment* for any rain that falls within the area. The upland ridge that separates one drainage basin from another is called the *watershed*.

River input

In heavy and prolonged rainfall, the total catchment area receives huge quantities of water. This is the *input*. Think of the volume of water produced by just

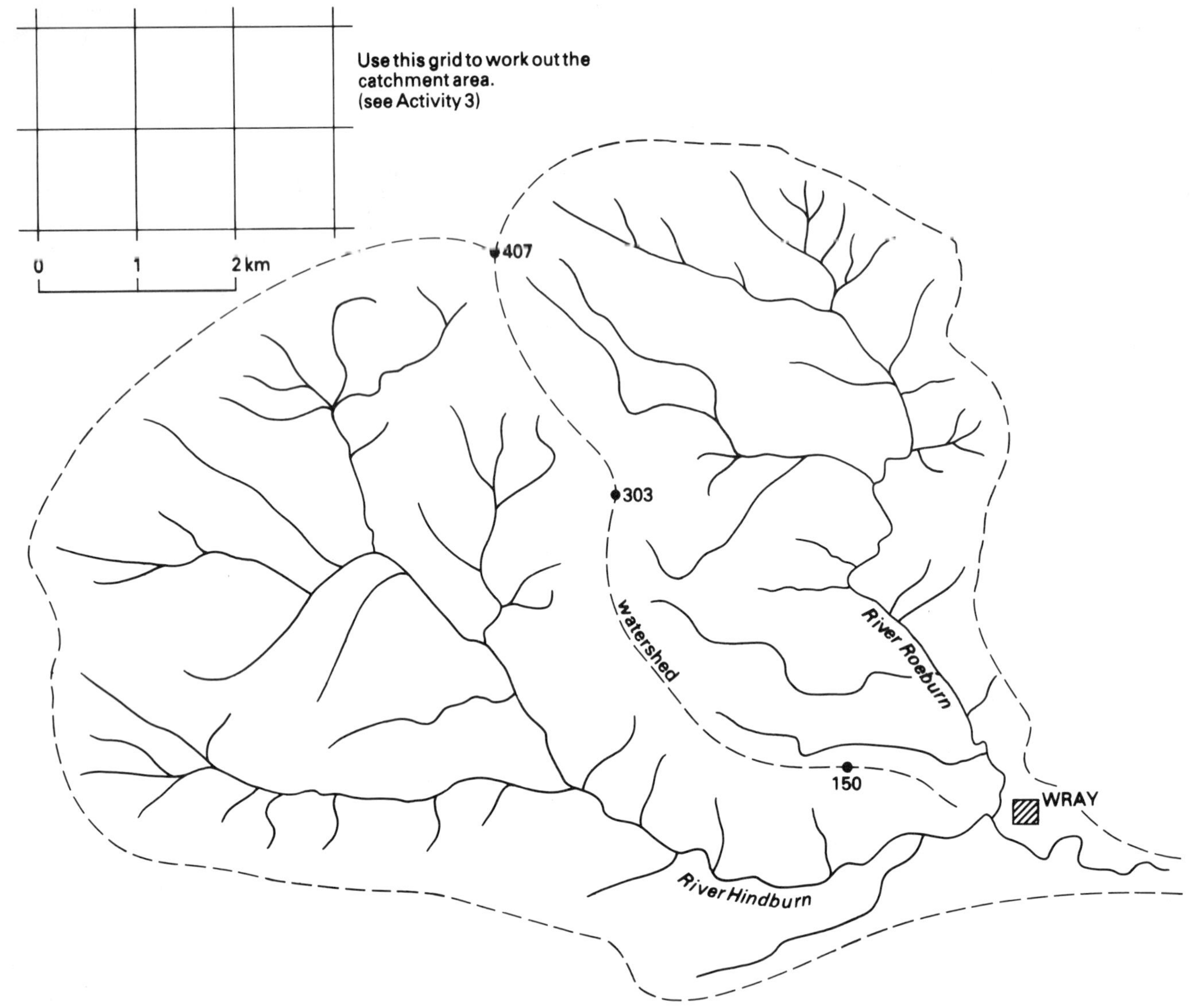

Fig 14.10 The drainage basin of the River Hindburn, with its main tributary, the River Roeburn. (Spot heights are in metres from sea level.)

Activity 3: How large is the catchment area?

1. Use tracing paper to make an area-grid with the same spacing as shown in Fig 14.10. Each square represents 1 km^2.
2. Overlay the grid on the map of the drainage basins. Count up the number of squares, allowing for half squares. How many square kilometres are there? This will give you the *catchment area* of the Roeburn and Hindburn river basins.

---

1 mm of rainfall! The volume of water is measured in cubic metres (m^3). Each square kilometre (km^2) receives 1 mm. Therefore, for 1 km^2:

$$\begin{aligned}\text{Volume of water } (m^3) &= \text{area} \times \text{rainfall}\\ &= 1\ km^2 \times 1\ mm\\ &= 1000 \times 1000 \times 0.001\\ &= 1000\ m^3\end{aligned}$$

Can you work out what the total input volume for the whole catchment area would be in this case? Your answer will be 1000 m^3 × the number of km^2 in the catchment area. See Activity 3.

Some of the input may well soak into the soil and surface peat. It may even reach the permeable sandstone beneath. The biggest danger of flooding is after several days of heavy rain. Why do you think this is? The permeable rocks, peat and soil will store a certain amount of water, but eventually they become fully saturated. The amount of run-off quickly increases. Then the water level in nearby streams and rivers rises very fast and flooding can occur.

River output

The water flowing out of the drainage basin is the *output* or *discharge*. In the Wray disaster, where 50 mm of rain fell in about an hour, the ground was already saturated. Therefore most of this input of water became surface run-off, and entered the streams. Using the calculations shown above, can you work out the volume in cubic metres?

A stream in flood has tremendous erosive power. For a few hours the volume of water passing through Wray was even greater than the normal total volume of the River Thames!

One local farmer did get some benefit from the disaster. A huge heavy boulder had blocked the entrance to one of his fields for many years. He had often unsuccessfully tried to move it. After the flood the boulder had gone, nor was it visible anywhere downstream from that point.

Activity 4: How does the load settle out?

River water is often cloudy when the stream is in flood.

1 Collect a sample of cloudy river water in a plastic bottle.
2 Allow the load to settle. The larger grains of gravel, grit and sand will settle within seconds. Finer silt particles stay in suspension much longer. Eventually all the load will settle in *graded layers*.
3 Measure the thickness of each graded layer. Make a labelled sketch of your results.
   a) Which layer is the thickest?
   b) How long does it take for the water to clear completely?
4 When the water is clear, evaporate a few drops of it in gentle heat on a watch glass.
   a) What do you see?
   b) What part of the river load is this?

Activity 5: Do pebbles move on a river bed?

On upland streams, the bedload of pebbles and boulders is easy to see, especially when the water level is low. One way to find out if they move is to mark a number of pebbles with paint. Then see if you can find them some time later. Do they move more quickly when the river is at full discharge?

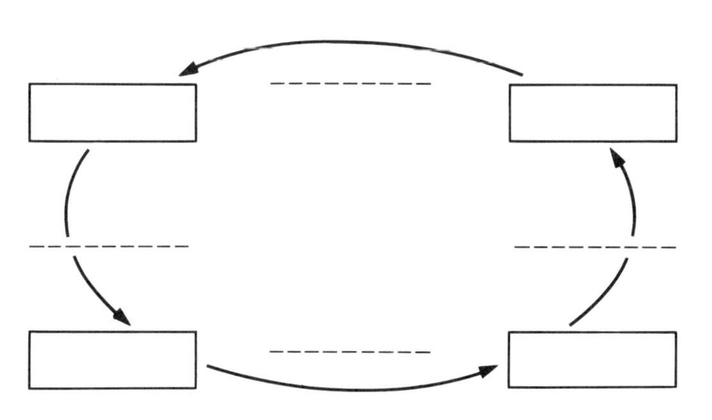

Fig 14.11 The water cycle

Exercise 1: The water cycle

(*Solution on page 144*)

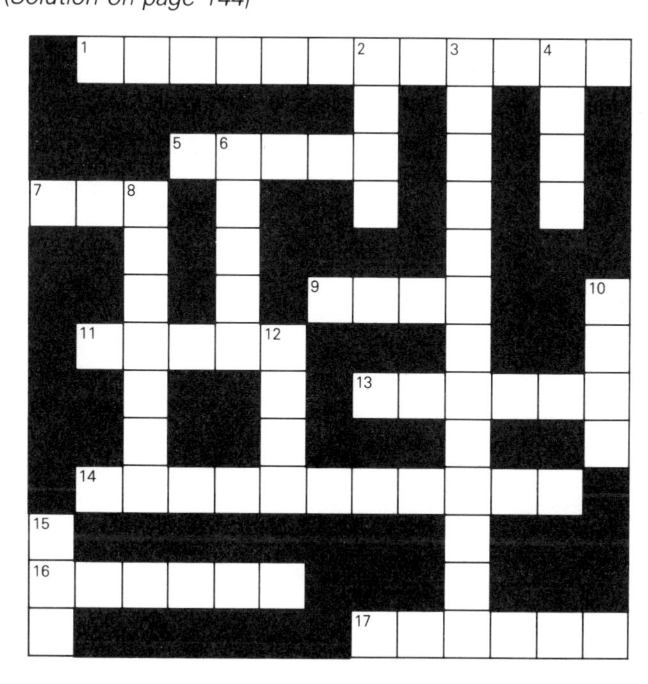

Across

1 The moisture found on cold windows; and up in the sky (12)
5 A large body of water (5)
7 Rivers flow into it (3)
9 Water will also ________ through the rocks (4)
11 A channel of running water (5)
13 Only rocks that are ________ can soak up water (6)
14 Rainwater soaks into rocks to become this (11)
16 The flow of water out of the ground (6)
17 The direct surface flow of water (6)

Down

2 Water will ________ into the soil (4)
3 Plants give out water vapour by this method (13)
4 Sometimes rain water will flow ________ the surface (4)
6 The name given to the circulatory movement of water (5)
8 A permeable rock that holds water (7)
10 Many rivers begin where water will ________ to the surface (5)
11 ________ is one kind of precipitation (4)
15 People make ________ of surface water (3)

Exercise 2

1 Make your own copy of Fig 14.11. Complete the diagram, using the words listed below. The words in capital letters are for the boxes.
   wind evaporation rain surface run-off CLOUDS CONDENSATION LAND SEA
2 Your completed diagram does not show *transpiration*, nor does it show *groundwater flow*. Improve it by adding this information.

3 Your home and your school have just been swept away in a massive flood! Write a report for a national newspaper. Include information on:
- damage caused by the flood
- eyewitness accounts
- raising money for relief funds afterwards, to help flood victims
- the disease threat from burst drains
- the reasons for the flood.

4 Write down the most accurate answer to each of the following sentences:
a) The area drained by a system of branch-like channels is a
A drainage basin
B catchment area
C river valley
D delta.
b) A watershed is a
A man-made lake
B point where two rivers meet
C ridge between two drainage basins
D movement of river water.
c) The basin discharge is
A overland flow
B input of water
C output of water
D channel flow.
d) Deposited river sediments are known as
A alluvium
B bedload
C suspended load
D debris.

5 a) Make your own labelled sketch of the photograph shown in Fig 14.7. Label the following features: bedload deposits; rebuilt house; embankment wall; direction of river flow; inside bend.
b) Is the river transporting any load in suspension?
c) Is the river transporting any bedload?
d) What kind of load is still being actively moved by the river?

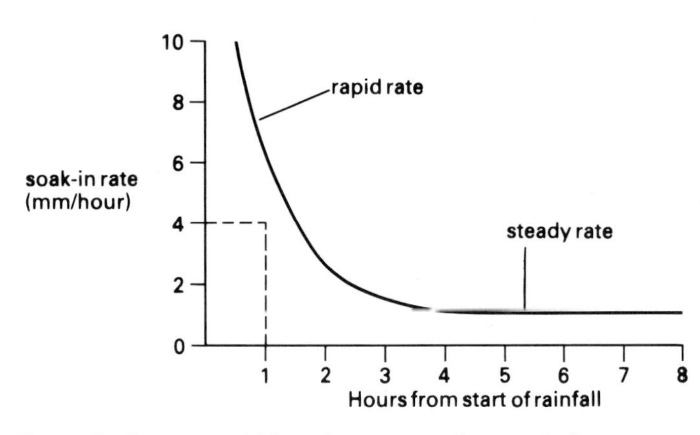

Fig 14.12 The rate at which surface water soaks away during a storm

Exercise 3

1 Study Fig 14.3. Describe all the factors that you can see that have led to the siting of the farms and villages.

2 Fig 14.12 shows the rate at which water soaked into the soil, during a storm of constant rainfall.
a) Why is the rate so rapid in the first hour?
b) Explain why it slows down to a steady rate later on.

3 a) Using graph paper, draw a graph as shown in Fig 14.12. Plot the following rainfall figures. Make a bar graph for each hour.

Hour	Rainfall (mm)
1	4 (already plotted)
2	3
3	6
4	4
5	7
6	3

b) What is the total amount of rainfall which did not soak into the ground?
c) What would happen to this excess water?

4 a) Explain fully what is meant by an artesian basin.
b) Why does water gush out of an artesian well?

5 Explain fully the meaning of the terms:
a) porosity;
b) permeability.

6 Explain fully how rivers use energy to erode and transport material.

15 BRITAIN: PAST AND PRESENT

The Earth's surface is in a state of constant change. This is partly due to the movement of the plates causing continents to move apart to create new ocean areas. Over millions of years, oceans may disappear as continents collide and fold mountain ranges are uplifted. Environments also change because of the intensity of weathering and erosion. For these reasons the Earth's surface environments cannot remain the same for any great length of time.

Layers of history

All the evidence for reconstructing the sequence of events in a particular area comes from a study of the rocks. Igneous and metamorphic rocks give useful clues about the past positions of the plate margins and fold mountain chains. Sedimentary layers, or *strata*, together with their fossils, provide the best evidence of past environments. The study of this evidence is known as *stratigraphy*. In stratigraphy it is normally assumed that if one layer or stratum lies on top of another, then the upper layer is the younger. This is known as the *Rule of Superposition*. The rule does not, of course, hold true if beds have been overturned in folding.

In many sequences of sedimentary rocks, each layer lies regularly one upon another. This indicates steady and regular deposition over a very long time. A sequence of this nature is said to be *conformable*. Sometimes a sequence of layers may be tilted or folded. The folded rocks may be uplifted and exposed to erosion, or they may subside and be buried under new layers of sediment before being finally uplifted and eroded. This series of events produces a feature known as an *unconformity*. An unconformity is evidence of a number of events that took place over a considerable period of time (Fig 15.1).

Fig 15.1 An unconformity at Portishead, Avon. A Triassic angular boulder bed rests on top of Devonian sandstones. This represents a break in deposition of 120 million years!

Table 1: Summary of the relationship of environment to scenery and facies

Environment	*Scenery*	*Possible facies*
Glacial	U-shaped valleys, aretes, moraines	boulder clay (till) – angular boulders with fine clay
Land volcanic	conical hills, lava plateaux	bombs (lapilli), welded tuffs, ropy and blocky lava flows
Submarine volcanic	submarine mountains	pillow lavas
Rivers	V-shaped valleys, meanders, floodplains	sand and pebbles deposited as sand and gravel beds. Slight rounding of grains but still angular
Deltas	flat river lowland extending out to sea, liable to flooding	sandstone with ripple marks and current bedding
Estuaries	wide tidal river mouth	mainly clays and mudstone
Deserts	rocky and sandy desert, with sand dunes	very well-rounded sand grains coated with iron oxide. Large scale cross-bedding.
Desert lakes	salt flats that flood on occasions	evaporite salt deposits, eg, gypsum/halite
Shorelines	eroding cliffs, stacks and arches, beaches, spits and bars	conglomerates, with well rounded pebbles, and ripple-marked sandstones
Shelf seas	sunlit shallow water, reefs common in tropics	sandstones, mudstones, and limestones. Marine fossils common
Offshore deeps	submarine canyons, turbidity currents	slumped sandstones and mudstones, very fine organic oozes

The importance of present-day environments

Another important aspect of the understanding of past environments is the study of *present* environments of erosion and deposition, together with the animals and plants associated with them (see Table 1). Many of the features seen in rock layers that are millions of years old, can be recognised in modern sediments. For instance, the rounded pebbles in a conglomerate indicate a former beach deposit on an ancient shoreline. Ripple marks on the bedding planes of sandstone indicate shallow water conditions and current bedding indicates a deltaic environment.

THE FACIES CONCEPT

The total character of a rock-layer is known as its *facies*. This includes the mineral content; the shape and size of the rock grains; the fossils it contains; and structures such as layering, ripple marks or graded and current bedding. A rock deposited in a particular environment will have its own distinctive facies.

Some facies can give precise detail on the climate and the temperature of the environment in which they formed, eg, oolitic limestone indicates a shallow tropical marine environment (see page 101). Fossils also provide very valuable evidence. Modern plants and animals are adapted to live in particular conditions. This was also true of animals and plants in the past. For example, sea urchins are only found today in shallow marine environments. They cannot tolerate fresh or even brackish water. The same is true of reef-building corals. It is safe to assume that fossil sea urchins and corals living millions of years ago were adapted to similar conditions. Careful analysis of fossil communities can give information on factors such as speed of water currents, depth and temperature. When this kind of information is gathered from a number of regions and combined it becomes possible to reconstruct the geography of the past, and to decide on where ancient shorelines might have been. A *palaeogeographic* map can then be drawn to summarise this information.

How to interpret a set of rock layers

A composite cliff face gives a cross-section as shown in Fig 15.2. The red sandstone is made up of large scale cross beds. These were formed when sand dunes were deposited on top of each other. The red tinge is due to the oxidation of iron – a reaction that takes place in dry conditions. The sand grains are rounded and well sorted. This suggests that they have been moved around a lot by the wind. Deposits of carbonate (calcite) are found in cracks and bedding planes in the sandstone. Today, similar *calcrete* (carbonate) deposits form a hard surface crust on soils in semi-arid regions. The conglomerate is *basal*, it represents an ancient beach deposit laid down by an advancing sea.

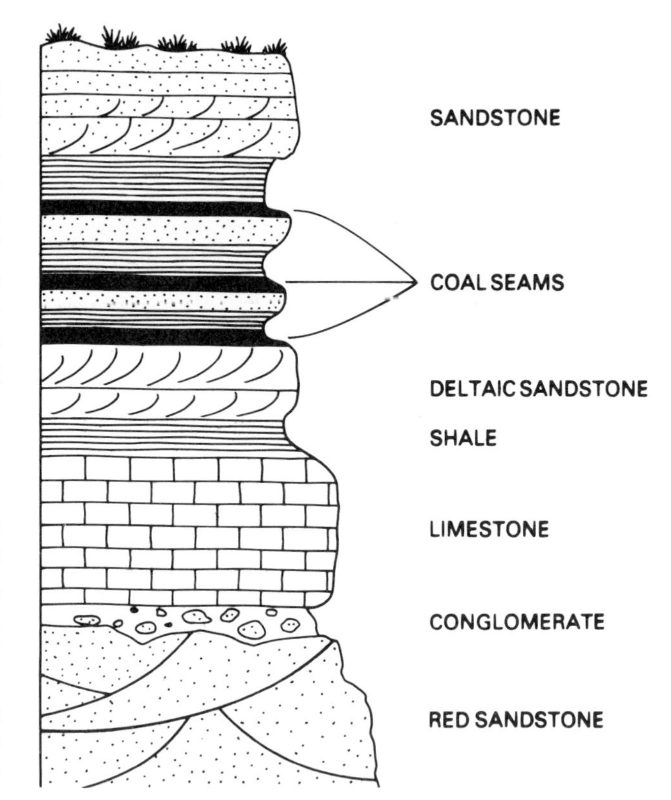

Fig 15.2 Section through a cliff-face

The limestone above the sandstone contains a large number of fossil fragments, showing that deposition was slow. Many of the fossils are broken and this suggests water currents. Most of the fossils are crinoids. This indicates a shallow, shelf sea environment. The presence of ooliths also confirms this theory.

The shale represents the deposition of fine mud particles. The sediment was possibly laid down offshore from a nearby river mouth. The shale is followed by a bed of sandstone. The sandstone is current-bedded. It is therefore likely to be deltaic in origin. The direction of the current beds gives the direction of river flow.

The coal seam above the sandstone shows that the top of the delta must have become overgrown with thick forested swamps. Organic matter fell into the swamp to form peat, and eventually coal.

The sequence of the repeated rock bands is typical of a delta advancing and retreating in an area of shallow sea (see page 116).

CHANGING ENVIRONMENTS IN THE BRITISH ISLES THROUGH GEOLOGICAL TIME

Latitude and climate

The movements of the Earth's plates have caused Britain to drift slowly northward. Britain has crossed the globe from somewhere near the South Pole in the Precambrian to its latitude position today of 60° N.

Using *palaeomagnetic* evidence, it has been possible to find out the past latitude position of Britain on the globe, at any particular time. Certain iron-bearing minerals in igneous rocks, align themselves in the Earth's magnetic field at the time of their formation. This information is 'frozen in' and retained when the magma cools and the rock solidifies. The dip of the tiny 'magnets' tells us the past latitude position (Fig 1.5).

The Precambrian

The oldest Precambrian rock is found along the coast of north-west Scotland and in the Outer Hebrides. It is known as *Lewisian gneiss*. This grey, crystalline, metamorphic rock was once volcanic ash and sand, deposited around a volcanic island chain 3000 million years ago. Today it forms a rugged terrain (Fig 15.3).

Most Precambrian rocks are very difficult to interpret. This is because the original sedimentary structures have often been lost by repeated intense folding, metamorphism and intrusion of magma.

The Lower Palaeozoic era
(Cambrian, Ordovician and Silurian)

When the Lower Palaeozoic era began 570 million years ago a wide ocean existed between northern and southern Britain. Shelf-sea limestones indicate that there were shallow marine conditions parallel to the shorelines (Fig 15.4). Marine life evolved separately on either side of the deep central ocean. For instance, Cambrian trilobites found in England and Wales are different types from those found in Scotland. Brachiopods and graptolites were also abundant at this time.

The wide ocean slowly began to close in the Ordovician period. The oceanic plates sank at subduction zones formed along both continental edges. The Scottish shoreline may have looked rather like the Andes mountain range does today. It took several million years for the ocean to close completely and the change was accompanied by a good deal of volcanic activity. Enormous quantities of andesite lavas were erupted in Snowdonia, the Lake District and along the northern edge of the Southern Uplands. Andesites are commonly erupted at subduction zones today. At the same time thick ocean sediments were intensely folded and uplifted to form the Caledonian mountain range.

By the end of the Lower Palaeozoic most of the British area was above sea level.

The Upper Palaeozoic era
(Devonian, Carboniferous and Permian)

During this era, Britain was drifting northward through the tropics. In the Devonian period, most of Britain became part of a new land area – the 'Old Red Continent'. Only the south of England was still under the sea. The shoreline lay roughly along a line joining London and Bristol. The new Caledonian mountains were rapidly eroded, producing conglomerates, sandstones and mudstones in the lowlands. To the south of the mountains there was a vast coastal plain. It was crossed by meandering rivers, depositing repeated layers of pebble-beds, sands and silts. The Devonian continental facies is known as the *Old Red Sandstone.*

In early Carboniferous times, a warm shallow sea

Fig 15.3 Scenery in north-west Scotland

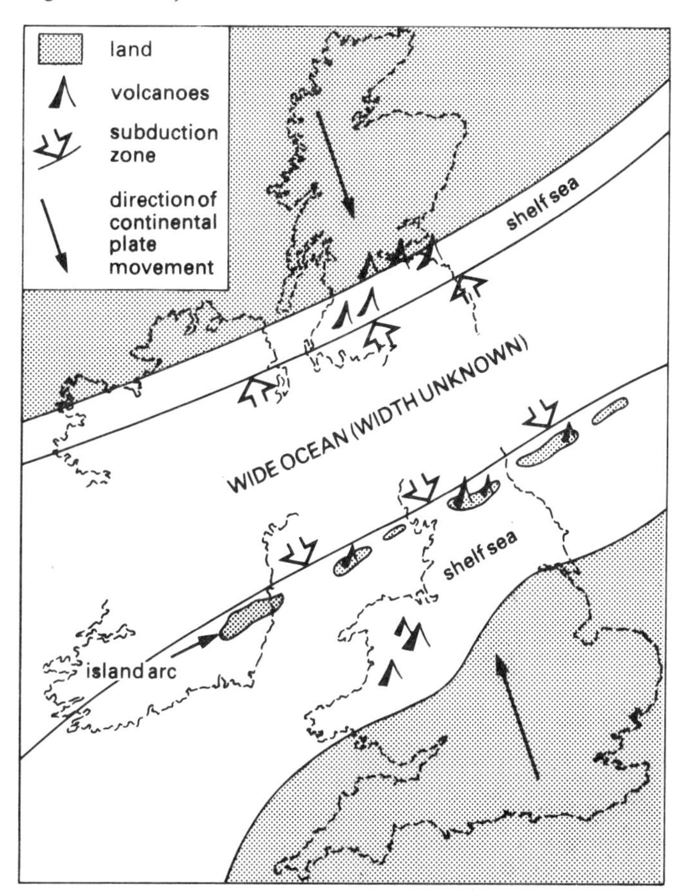

Fig 15.4 A palaeogeographic map of Britain in Ordovician times

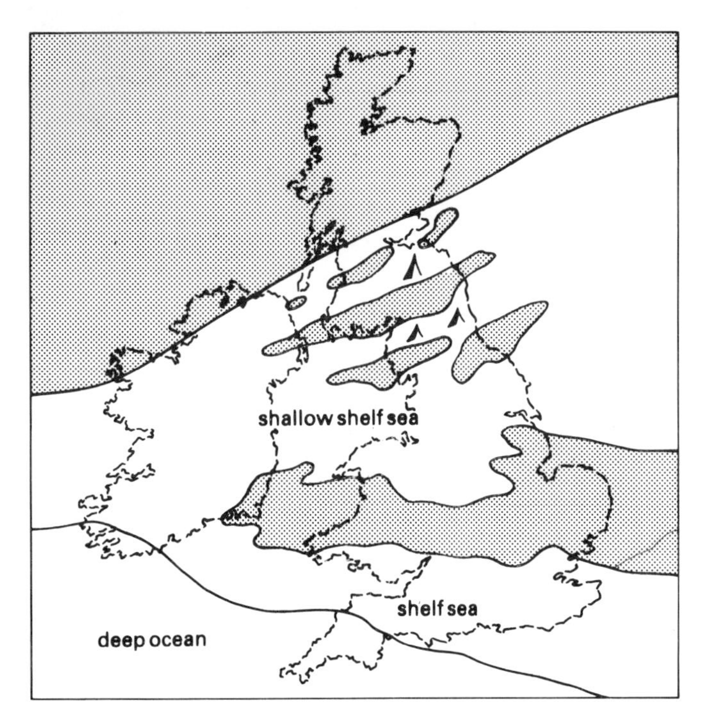

Fig 15.5 **Britain in Carboniferous times**

steadily advanced from the south, flooding the lowland areas as far north as the Scottish Highlands. Some large areas remained as islands (Fig 15.5). The most typical deposit at this time was *Carboniferous Limestone.* This limestone contains the stem fragments of crinoids, brachiopods and corals.

Eventually the sea began to retreat as large deltas were built out into the sea from the land. The deltaic deposits were mainly coarse sandstones or 'grit'. This facies is known as the *Millstone Grit.* Bands of marine shale among the sandstones show that the deltas were sometimes flooded by the sea. The tropical deltaic environment was ideal for the development of widespread coastal swamps and the *Coal Measures* were laid down towards the end of this period. Deposition was often in the form of repeated sequences known as *cyclothems.*

Each cyclothem is made up of shale, sandstone, and coal, repeated many times. It reflects the following events:

1 Invasion of the swamp by the sea, and deposition of marine muds (shale).
2 Return to a river delta and deposition of non-marine sands (sandstone).
3 Creation of the coastal swamp and accumulation of peat (coal).

The Devonian and Carboniferous sediments were intensely folded and faulted. The granite bathylith of the south-west peninsula was also intruded at this time. Further north there was more gentle folding and general uplift.

Towards the end of the Carboniferous times, red sandstones and shales were deposited in arid conditions and the coal swamps disappeared. Plate collisions to the south and west of Britain caused the uplift of a new range of fold mountains known as the Variscan (Hercynian) Range. The northern edge of this fold belt ran through south-west England and southern Ireland.

By Permian times the continents were grouped together to form the supercontinent, Pangaea. Britain was then about 20° north of the equator in the middle of a large continental mass, with a desert climate. This situation continued into the Triassic period. Permian and Triassic rocks consist of conglomerates and large scale cross-bedded red sandstone with well-rounded sand grains. This facies is known as the *'New Red Sandstone'.* Study of the dune (cross) bedding in the sandstones has shown that the wind blew from an easterly direction and that Britain lay in the trade wind belt, just north of the equator (Fig 15.6a/b). Evaporite salt beds were laid down in a shallow gulf inlet from the Permian Sea (Fig 15.7). Formation of evaporite deposits is explained on page 97.

Fig 15.6 (a) A sandy desert with dunes, Namibia (b) Cross bedding in desert sandstones

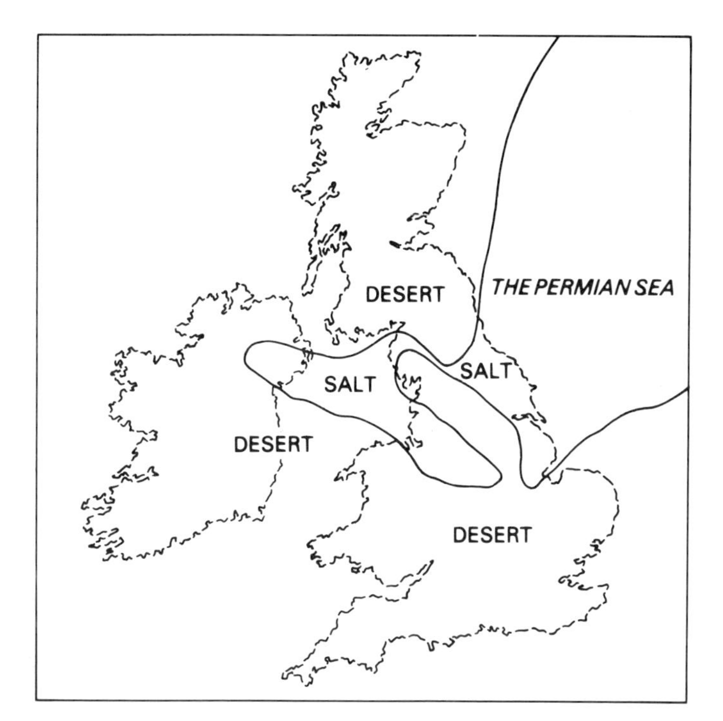

Fig 15.7 Britain in Permo-Triassic times. For 20 million years this shallow inlet of the Permian sea repeatedly evaporated and was then reflooded. The cycle of deposition of salts, shown in Figure 12.5, was repeated many times.

The Mesozoic era (Triassic, Jurassic, Cretaceous)

Desert conditions continued in the Triassic. So did the erosion of the mountain areas to the north and south. By the beginning of the Jurassic, the Pangaea supercontinent was beginning to break up. Most of the mountains in the British area had been worn down and the sea advanced (Fig 15.8). There was widespread deposition of alternating limestone and clay in shallow, warm, shelf seas. Ammonites and belemnites were abundant, as well as bivalves, brachiopods and corals. Land animals included the dinosaurs. Birds and the first flowering plants appeared (see Chapter 10).

In the early Cretaceous period, the Atlantic Ocean began to open. At the same time there was some uplift in the north-west of the British area. A sequence of limestones, mudstones and sandstones was laid down in the south-east of Britain. To begin with, this was largely a deltaic/freshwater environment. It was followed by a shallow water, marine environment as the sea level rose again.

Later in the Cretaceous the chalk beds were deposited (Fig 13.3). Chalk is a very pure, fine-grained limestone. There is very little sediment mixed with the tiny coccolith (algal) remains of which it is composed (see page 100). The chalk sea advanced to cover almost the whole of the British area. The nearest land was many miles away.

The Cainozoic Era (Tertiary and Quaternary)

The chalk seas ended when the crust was uplifted. This uplifting was due, in part, to the opening of the North Atlantic Ocean just to the north-west of Britain. The result was that most of the British area became land (Fig 15.9). At the same time, the Alpine mountain range was being uplifted as the Tethys ocean closed, due to the collision of the African and Eurasian plates. This caused slight secondary folding and tilting of earlier sedimentary rocks.

Igneous activity was very intense, especially in the early Tertiary. Basalt lavas were extruded in huge

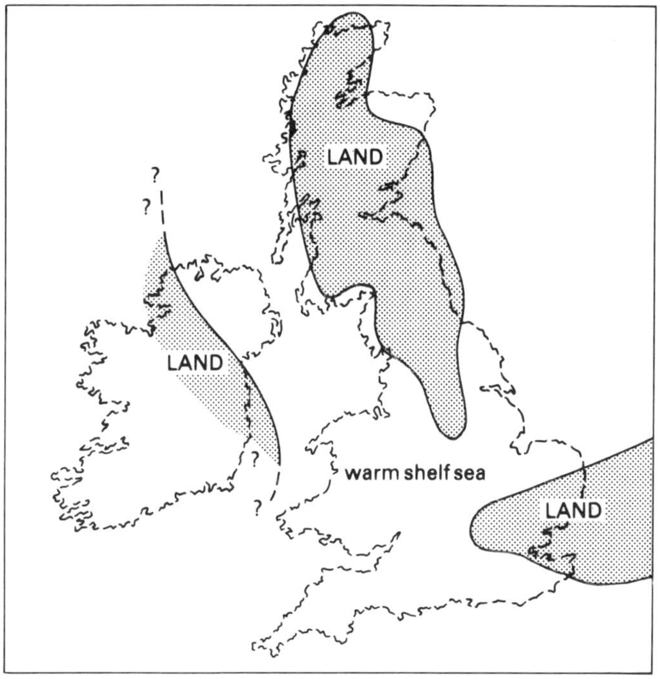

Fig 15.8 Britain during the Jurassic period

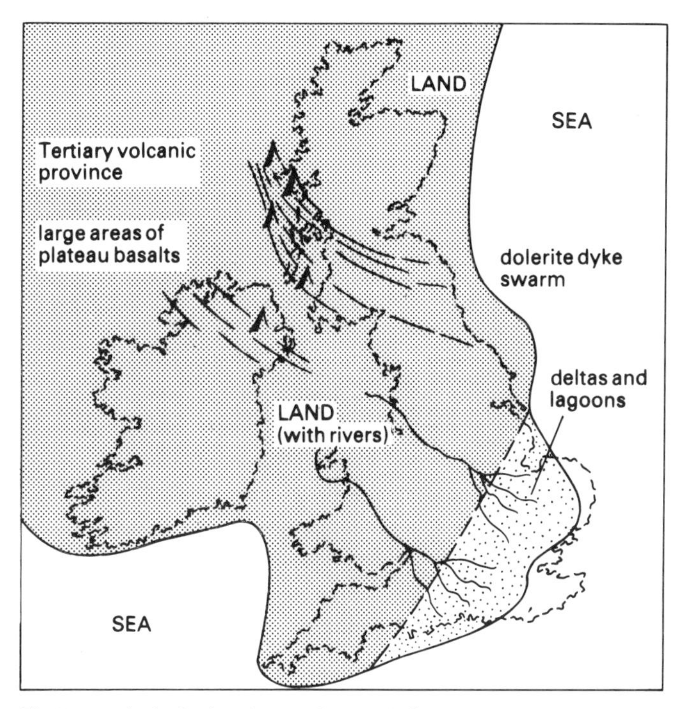

Fig 15.9 Britain during the Tertiary period

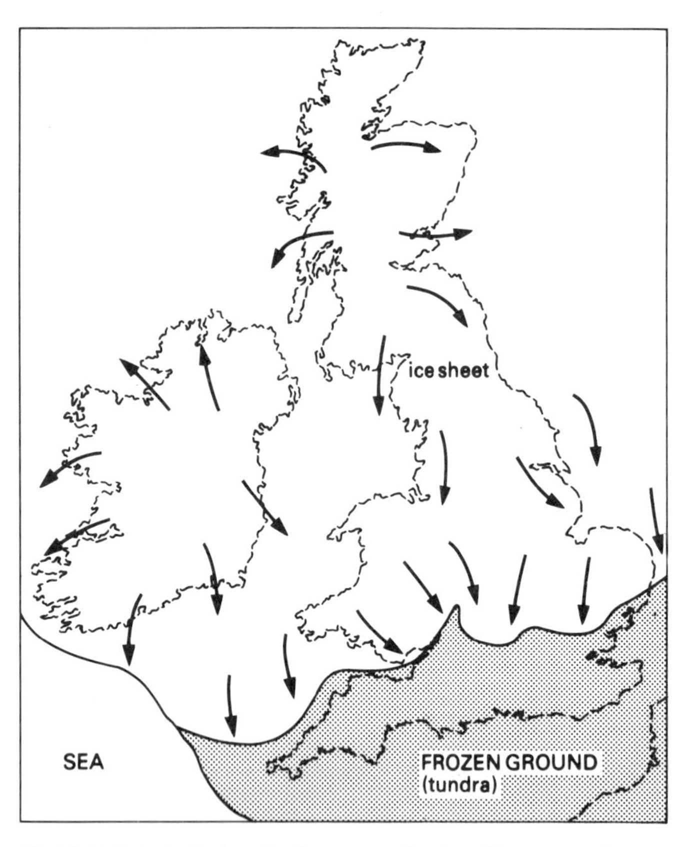

Fig 15.10 Britain during the Quaternary Ice Age. The arrows show the main directions of ice flow

Fig 15.11 The Barnard Glacier, Alaska, USA, showing lateral and medial moraines

Fig 15.12 A U-shaped valley

quantities from several volcanic centres. Dyke swarms were also intruded into surrounding rocks. The igneous activity is evidence of the formation of a constructive plate margin. The scenery of the north-west of the British Isles at that time was much like Iceland is today. However, soil layers preserved in between lava flows indicate a much warmer climate than Iceland has today. This agrees with the palaeomagnetic evidence for Britain's latitude position of 40° N.

The Quaternary Ice Age (began 2 million years ago)

By this period Britain had drifted to its present position. Fossils indicate a general cooling of world-wide climates as Britain and much of Europe became buried under thick ice sheets. The ice advanced and retreated many times. Each cooler period of advance is known as a *glacial*, and a warmer period of retreat as an *interglacial*. Fossils show that the climate in some of the interglacials was much warmer than it is today. The last glacial period in Britain ended about 12 000 years ago. We are now in an interglacial that could end at any time and plunge us back into an ice age (Fig 15.10).

Ice sheets form when snow builds up on high ground at a faster rate than it melts. It then compacts to form ice under the weight of new snow. Eventually the thick ice begins to slowly flow downhill. A moving mass of ice in a valley is called a *glacier* (Fig 15.11). Glaciers can gouge out huge quantities of rock and carve very deep valleys. Some of the Norwegian fjords are 1500 m deep with near-vertical sides.

BRITAIN TODAY

Glacial scenery

Much of the scenery in this country is the result of the work of ice. The shapes carved by valley glaciers can be seen in all the upland areas of the British Isles (Fig 15.12). A glacier enlarges and deepens a normal V-shaped upland river channel, to a U-shape. At the top of the valley is an ice-field supply area near the high mountain peaks. A *cirque* glacier may form on a mountain slope (Fig 15.13). The ice grinds downwards and rotates to form a hollow as it erodes backwards into the mountainside. When the ice melts, these hollows become small lakes or *tarns* (Fig 15.14). If cirque glaciers form on opposing slopes of a mountain-side, they both cut backwards. A knife-like ridge called an *arête* is then formed. If this happens on several sides a *horn peak* is formed, eg, the Matterhorn, in the Swiss Alps.

The bottom surface of a moving glacier is like a huge sheet of sandpaper. Rock fragments embedded in the moving ice scrape and scratch their way across the landscape. These scratches, known as *striations*, are

Fig 15.13 Cirque glaciers in the Alps

Fig 15.14 An arête, cirque and tarn – Helvellyn, Cumbria

visible on bare rock surfaces today. They indicate the direction of past ice movements (Fig 15.15).

Ice sheets and glaciers transport material in the form of *moraines* (rock debris). These are visible as dark areas on the glacier. Scree material falls onto the sides of a glacier to form *lateral moraines*. Where two glaciers merge, the lateral moraines combine to form a *medial moraine*. The end of the glacier is called the *snout*. Here the ice melts and deposits the material in a *terminal moraine*. A terminal moraine left by the furthest advance of British glaciers extends in a line across England from the Thames estuary in Essex towards Bristol. Material deposited in moraines is called *boulder clay* or *till*. It is a poorly-sorted mixture of jagged boulders mixed with fine clay. Boulder clay

Fig 15.15 Glacial striations on a bare rock surface

Fig 15.16 Glacial boulder clay or *tillite*

Fig 15.17 A glacial erratic block at Norber Austwick, North Yorkshire

compacts to form a hardened rock called *tillite*. Tillite deposits record past glaciations like the one in the southern hemisphere 300 million years ago (Fig 15.16).

Huge blocks of rock are often carried long distances before being dropped by the ice. These blocks are known as *erratics* and are not related to local underlying rocks. Distinctive erratics can be matched with their source areas. For example, Shap granite from an intrusion in north-west England can be found in erratics on the Yorkshire coast near Scarborough. From this it is possible to work out the direction of ice flow (Fig 15.17).

The shaping of the scenery

The detail of our present scenery has been shaped within the last half million years. The basic features of Britain's highlands and lowlands had been formed before that by uplift and tilting during the Tertiary. The highlands of Scotland, Northern England, Wales and Ireland have been *reactivated*. This means that the old 'roots' of the worn-down Caledonian and Variscan fold mountains were uplifted again. Britain's highlands are mostly formed of ancient crystalline rocks and tend to resist erosion.

The lowlands in England are formed by the softer, younger sedimentary rocks, largely tilted towards the south-east. The harder layers of limestone, chalk and sandstone resist erosion. They form *escarpments* (upland ridges) with the softer clay layers eroded into broad vales (Fig 15.18).

Coastal scenery

The British coastline is undergoing constant change. Certain coastal sections are being eroded by wave action. This means that the sea is advancing. In other sections of the coast the sea is depositing material and retreating.

Wave action undercuts the cliffs. This may cause the cliff to collapse. Sometimes waves will carve a cave at a point of weakness in the cliff, for example, a joint or fault line. If caves form on either side of a *headland*, they may be cut back until they join to form a natural *arch*. When the arch collapses, a *stack* is left as an isolated feature (Fig 8.15).

Where the coastline is composed of alternate hard and soft bands of rock, the waves will erode the softer rock at a faster rate. This is known as *differential erosion*. If the strata lie at right-angles to the coastline, the harder bands will form *headlands* and the softer bands will be eroded more quickly to form *bays*. When the 'grain' of the rocks is parallel to the coast, the sea may break through the more resistant rock layers. A *cove* is then hollowed out by wave-erosion in the softer layer behind it, eg, Lulworth Cove, Dorset (Fig 15.19), and Worbarrow Bay, Dorset.

Erosion produces great amounts of material. Pebbles,

Fig 15.18 Two escarpments caused by tilting of layers that resist erosion – Wenlock Edge (*left*) and Aymestry limestone escarpment (*right*)

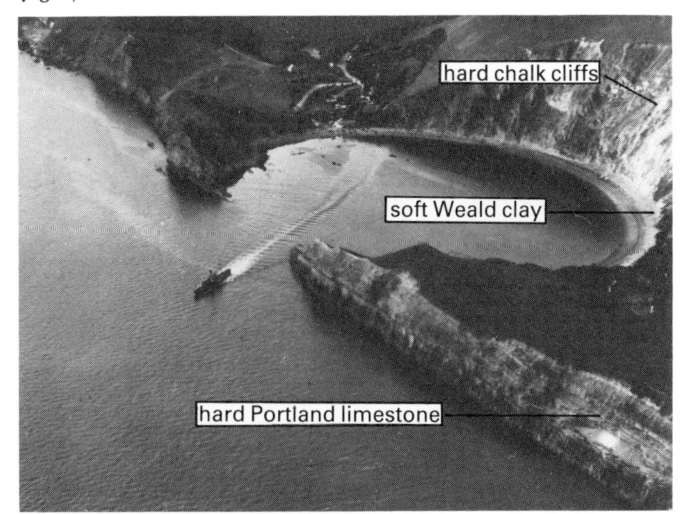

Fig 15.19 Lulworth Cove, Dorset

Fig 15.20 Chesil Beach, Dorset

Fig 15.21 Drowned valley or *ria*, Salcombe, Devon

sand and mud are frequently transported and deposited by wave action and form *beaches*. When the wave direction approaches the shoreline at an angle, loose material is transported along the beach. This process is called *longshore drift*. The material may be built out to sea as a *spit*, a low narrow ridge of sand or pebbles. Sometimes a spit can extend across a bay and join up with the coastline again to form a feature called *a bar*. Chesil Beach, near Weymouth, is one such example (Fig 15.20).

Many features of river valleys and the coast are formed by changes in land and sea level. During the last Ice Age glacial period, sea level was lower than it is today, so low in fact that Britain was joined to the mainland of Europe. Large amounts of water were locked up in the extensive ice sheets. The rivers of southern England cut deep valleys to the lower sea level at this time. When the climate improved and the ice melted, sea level suddenly rose and flooded many valleys.

A drowned river valley is called a *ria* (Fig 15.21). They are common in south-west Ireland and south-west England. A drowned glacial valley is called a *fjord*. Fjords are common around the coastlines of western Scotland and Ireland, eg, Loch Linnhe, West Scotland.

Recent changes

When the thick ice sheets that had covered Britain melted, the heavy load that had been depressing the crust was removed. Slowly the land began rising. In places it is still rising by a few millimetres each year. This is known as *isostatic uplift* or *rebound* (p.6). This process caused the uplift of wave-cut shorelines.

Fig 15.22 Raised beaches on the island of Islay, Argyll

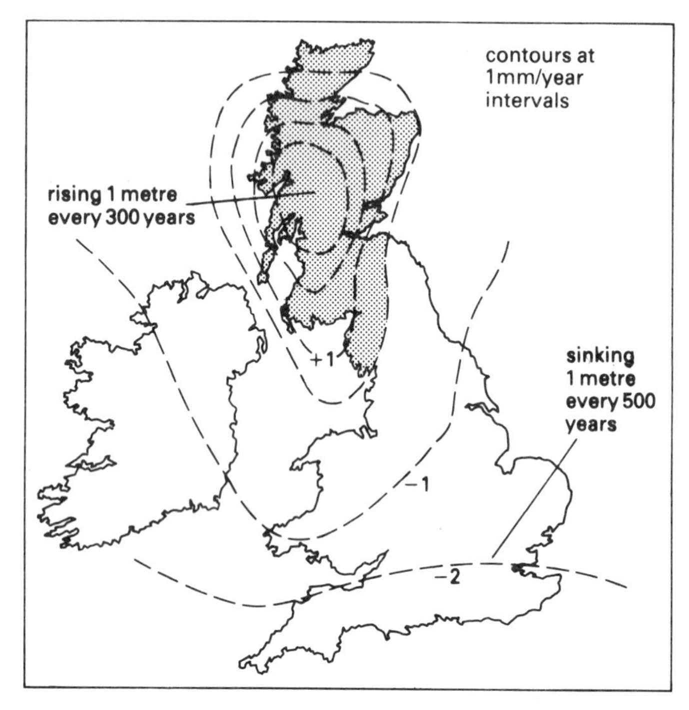

Fig 15.23 'See-saw' Britain. Isostatic rebound is causing uplift in the north, but in the south the sea is advancing. (Contours at 1mm/year intervals)

They became *raised beaches* far out of reach of the highest tides today (Fig 15.22).

The British Isles are still changing shape because of active erosion. The sea is advancing, especially in the south of the country as the land slowly sinks. That is why the threat of extensive flooding in places like London is becoming steadily greater. London is slowly sinking at the rate of 1 metre every 500 years. On the other hand, Scotland is rising 1 metre every 300 years (Fig 15.23).

Exercise 1

Make your own word square puzzle. Use any of the *italicised* words in this chapter. Test the puzzle on your friends.

1 Begin by placing tracing paper over the word-square grid (page 50).
2 Fill in as many words as possible. They may be vertical, horizontal, diagonal, or even backwards.
3 Write down the words you have used in a list below the puzzle.
4 Finally, fill in any blank spaces with unusable letters.

Exercise 2

1 Complete each statement with the most accurate phrase.

a) Stratigraphy is
   A a kind of geography
   B a study of sedimentary layers
   C a study of past events
   D the layering of sedimentary rocks

b) An unconformity is
   A a younger horizontal series of rocks
   B a surface between rock units that marks a gap in the geological record
   C a series of rocks deposited in a delta
   D a younger rock deposited on top of an older rock layer

c) A facies is
   A the alignment of mineral grains in a rock
   B movement along a fault line
   C the distinctive fossils, minerals and structures, in a portion of a rock unit
   D a particular environment of deposition

d) A cyclothem is
   A a repeated sequence of sedimentary layers
   B recycling of rock grains by erosion and deposition
   C the formation of evaporite deposits
   D the formation of peat in a forest swamp

e) A cirque is
   A a rounded cone-shaped hill
   B the moraine material left by an ice sheet
   C a glacier that hollows backwards into the mountain-side
   D a knife-like glacial ridge

f) An erratic is
   A a poorly-sorted mixture of clay and boulders
   B an alien boulder that is not related to the underlying rocks
   C rock material carried long distances by a glacier
   D a mound of loose rock left at the snout of a glacier

g) Isostatic uplift is
   A a change in sea level
   B upward movement of the land to compensate for the removal of mass
   C the melting of thick ice sheets that depress the crust by their mass
   D the formation of raised beaches

Exercise 3

1 Define what is meant by the Rule of Superposition. What exceptions are there to this rule?
2 a) Make a sketch of the unconformity (Fig 15.1).
  b) What kind of rock would you expect the boulders to be in the Triassic boulder bed?
  c) Explain fully the significance of an unconformity.
3 a) What is meant by the term 'facies'?
  b) Explain, with examples, how different facies can give important detail on the past environment of formation.
4 What is meant by a 'welded tuff'? What kind of environment of deposition does this rock indicate? (Also see Chapter 6)
5 What kind of facies would you expect to be produced in a tropical shelf sea?
6 Explain fully the effects of the Ice Age on British scenery.
7 Make a labelled sketch of Lulworth Cove, Dorset (Fig 15.19). Explain the processes that have formed this feature.
8 Describe a coastal feature that
  a) indicates the land is rising
  b) indicates the land is sinking.

J

16 USING ROCKS

Fig 16.1 Houses in St Ives, Cornwall

The rocks that make up Devon and Cornwall not only provide striking scenery, they are also an important source of employment and wealth. The main theme of this chapter is to show how the rocks in one small area are used in many different ways.

THE ROCKS OF DEVON AND CORNWALL

Granite and slate

Granite is an igneous rock. It formed as a huge intrusion into the rocks below Devon and Cornwall. This happened 300 million years ago. There was intense folding and pressure at that time, changing soft shales into hard slates. The heat from the intrusion helped to metamorphose the surrounding shales. The zone around a granite intrusion is called a *metamorphic aureole* (p.61). Here the changes are at their greatest. Erosion has since exposed the granite batholith in the form of granite moorlands, such as Dartmoor (see p.40). Much of the lowlands are underlain by slates and other metamorphic rocks.

The scenery in much of Devon and Cornwall is of granite uplands, such as Dartmoor. Sheltering around the edges of the moors are many towns and villages. It is easier to live in the lowlands where the land is sheltered, and communications are better. Traditional building makes use of local materials, for example, granite and slate (Fig 16.1). The slate which is quarried in Cornwall, splits easily into thin roofing slates. Granite is a very strong, hard rock. It has long been used for building houses and bridges throughout Britain. Granite chippings are also used as roadstone.

Mineral veins

As the granite cooled, hot metallic solutions moved upwards through mineral veins in the surrounding rock. The watery solutions carried many dissolved minerals. A number of important metallic ores were deposited. Their crystals line the walls of each vein.

Metallic minerals crystallize out at different temperatures. *Tin* is found nearest to, or even within, the granite mass, where the temperature was highest. *Copper* is found farther out. *Lead, zinc* and a little *silver* are found in the coolest part of the vein. They are farthest from the granite mass. By far the most economically valuable metal ore is tin (Fig 16.2).

Tin

Cornwall has long been a valuable source of tin. The Romans were first attracted to Britain because of this resource. Today Cornwall produces tin worth £30

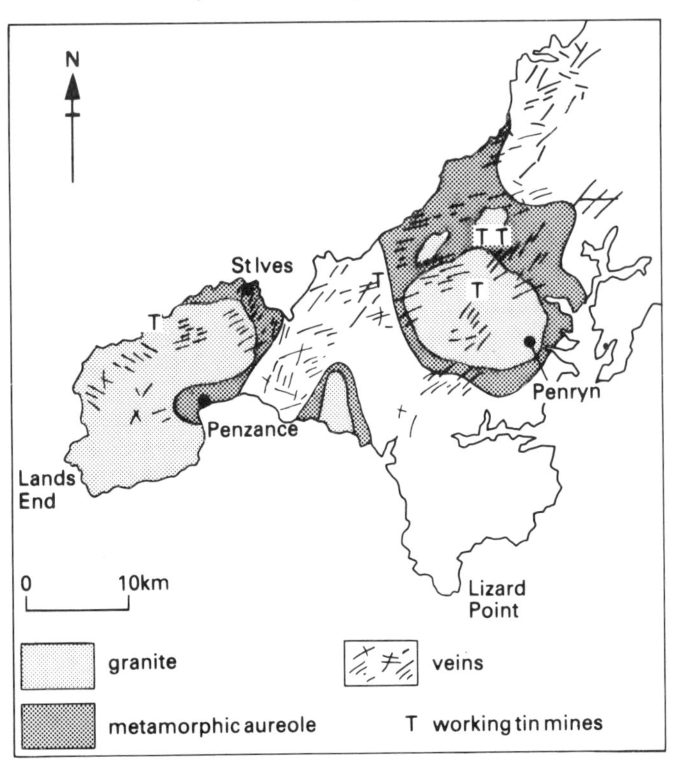

Fig 16.2 The rocks of western Cornwall, showing the granite intrusions and mineral veins

Fig 16.3 The new and the old: South Crofty tin mine. Modern large-scale workings such as these contrast with the many small tin mines of the past.

million a year. This provides Britain with about one-third of its needs. The tin industry was in decline in the twentieth century until world tin prices improved in the 1970s. Now there are eight working tin mines. Most of the tin comes from the South Crofty Mine. This is the largest tin mine in Europe (Fig 16.3); 300 people work at depths of up to 670 metres. The ore comes from 16 different veins and 21 000 tonnes are produced every month. Zinc, copper, silver and tungsten are important by-products.

Uses (Table 1)

Tin is expensive compared to other metals but it is very useful to industry. It is resistant to corrosion and it mixes easily with other metals. Tin is mainly used for making tinplate – a very thin layer of tin is deposited on a moving sheet of steel. Tinplate is used to make tin cans because it is much more economical than using pure tin.

Table 1 Main uses of tin

Use		
Tinplate *(approx 55%)*		for tin cans
Alloys *(approx 25%)*		tin and lead = solder tin, lead and antimony = pewter tin and copper = bronze tin and zinc = brass
Bearings	*(approx 20%)*	used in the making of low-friction bearings and pumps
Glass		for plate glass manufacture
Other uses		in paint, disinfectant, fungicide, wood preservative, cast iron

A quarter of all tin produced in Britain is mixed with lead to make *solder*, for welding purposes. Tin is also used in the glass industry. Highly-polished plate window glass is made by floating molten glass on a

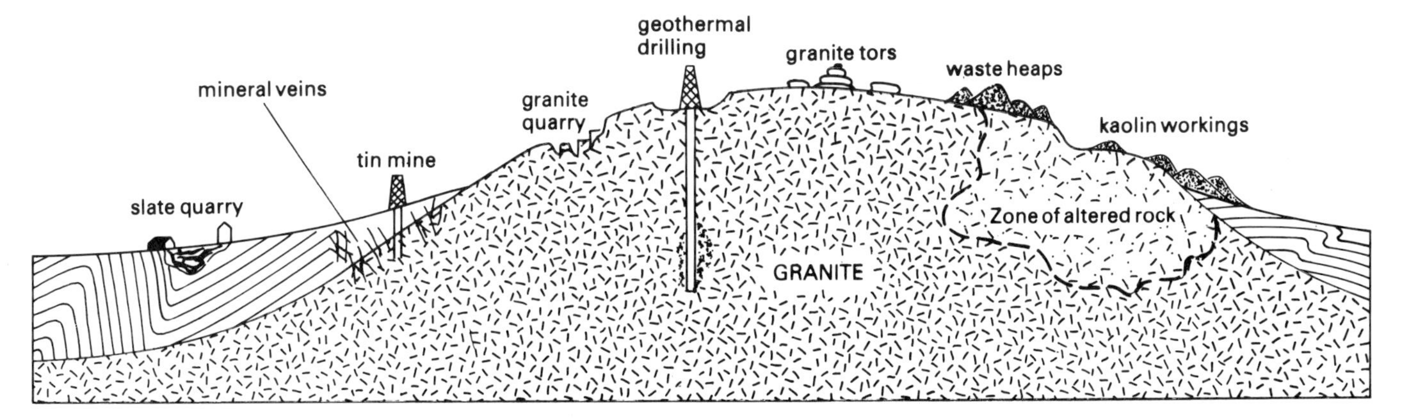

Fig 16.4 Cross-section of a Cornish upland, showing the different ways in which the rocks are used

bath of molten tin. The tin provides a perfectly smooth surface for the floating glass. The glass is allowed to cool and thus a perfect polished sheet of glass is formed.

Kaolin (china clay)

There are many *kaolin* pits in Devon and Cornwall. The quarrying of the clay takes place at the edges of the granite moor (Fig 16.4). Kaolin is formed by the alteration of granite. It is a pure, white clay. Granite pushed its way into the surrounding rocks, then hot acidic gases, such as carbon dioxide, were forced up from below. The gases rotted parts of the granite and the feldspar minerals in the rock were altered to a soft, white clay.

The china clay is washed from the rock face, using high pressure water hoses. The liquid clay is refined to remove unwanted mica and quartz. Then it is dried to produce a putty-like clay.

Uses (Table 2)

Table 2 Main uses of kaolin

	%
Paper manufacture	78
Ceramics	14
Paint, cosmetics, medicine and plastics	8
	100

At one time kaolin was mainly used to make bone china, ie, porcelain. Today its main use is in making paper. It is added to the wood pulp to fill the small spaces betwen the wood fibres. Glossy magazine paper is made of up to 30% kaolin. The higher the percentage of kaolin the glossier the paper.

The finest china is still made from pure kaolin. A range of ceramic products are made from the less pure kaolin. These products include earthenware, bathroom fittings and roofing tiles. Heat-resistant ceramics are being developed to replace metal parts inside high-temperature engines. Ceramic tiles are used on the American space-shuttle to protect it from the heat of re-entry.

The environment

A common eyesore in parts of Cornwall and Devon are the 'white mountains'. These ugly mounds of dazzling white sand-waste are produced by the china clay industry. Recently a use was found for this sand-waste and it is now used to make concrete blocks and bricks. It may also have other uses in the building industry.

If the mountains of sand-waste could be removed and used, the environment of Cornwall and Devon would be improved. The beautiful scenery attracts many tourists every year. Tourism is an important source of wealth to the area and could be lost if no attempt is made to clear up the mess from mining (Fig 16.5).

Fig 16.5 A kaolin quarry near St Austell, Cornwall, with unsightly spoil heaps

GEOTHERMAL ENERGY

Geothermal energy is the heat contained in rocks beneath the surface. In volcanic areas the heat is very great. In Iceland and New Zealand the water from hot springs is piped directly to houses to heat them. The steam is used to make electricity.

In Britain there are no active volcanoes, but the rocks in some places are hotter than in others. It is normal for temperature to increase with depth. Hot water at 70 °C has recently been found underground at Marchwood, in Hampshire, this water can be piped and used to heat houses and public buildings. It is not hot enough to make electricity. The Romans discovered the hot springs at Bath. The temperature, 46.5 °C at the surface, is just right for bathing. This water fell as rain on the Mendip Hills, 15 kilometres away. It takes 2000 years for the water to seep down through the permeable limestone (Fig 16.6). At a depth of two and a half kilometres the water is heated to 80 °C.

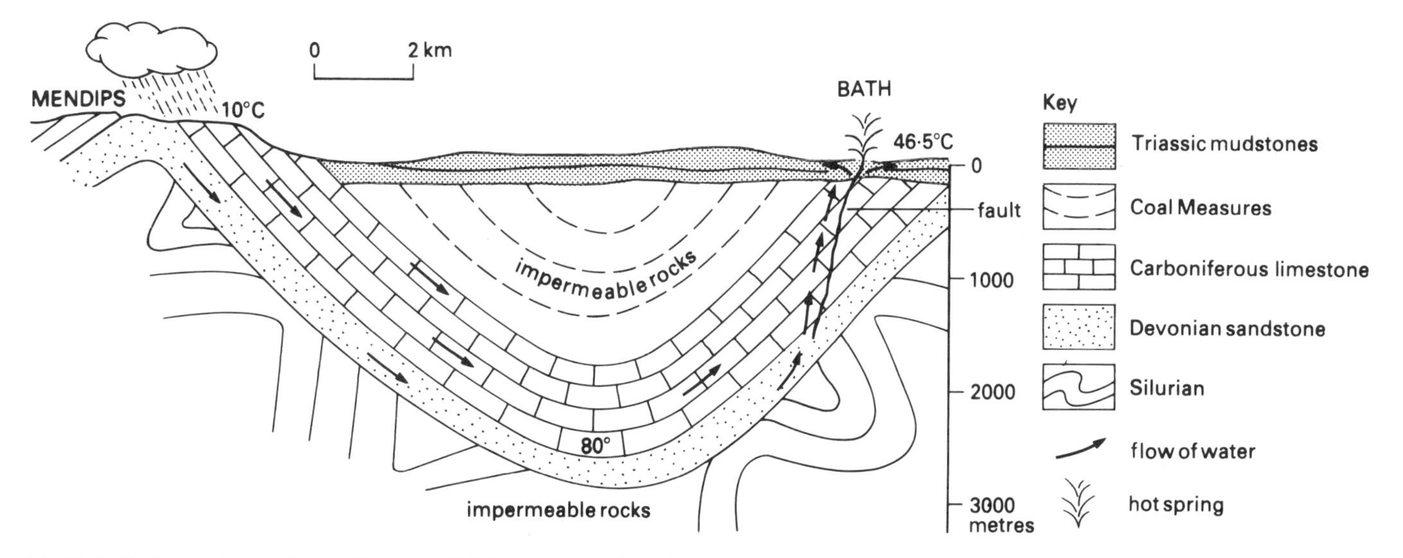

Fig 16.6 The hot springs at Bath. The permeable limestone and sandstone act like a huge pipe, and allow water to pass through them. They are sandwiched between impermeable rocks, above and below. Water is heated at depth. It then flows out along the faultline at Bath under pressure. The pressure is due to the greater height of the Mendip Hills, compared to Bath.

The hot rock project

In a disused granite quarry at Penryn, in Cornwall, an attempt is being made to extract the heat from the granite beneath. Granite contains some radioactive minerals. These minerals produce heat as they decay. This makes granite hotter than most rocks but it is only noticeable deep underground. At a depth of 2000 metres the temperature is over 100 °C.

There is no natural spring of hot water at Penryn. Instead hot water has to be pumped to the surface. Two boreholes are drilled, 10 metres apart, to a depth of 2000 metres. Explosives are used to break up the rock around the lower borehole. This creates a system of small cracks, leading to the upper borehole. Cold water is then pumped down the lower borehole. The granite heats the water as it seeps through the fractures. Then the water is pumped to the surface through the second borehole (Fig 16.7).

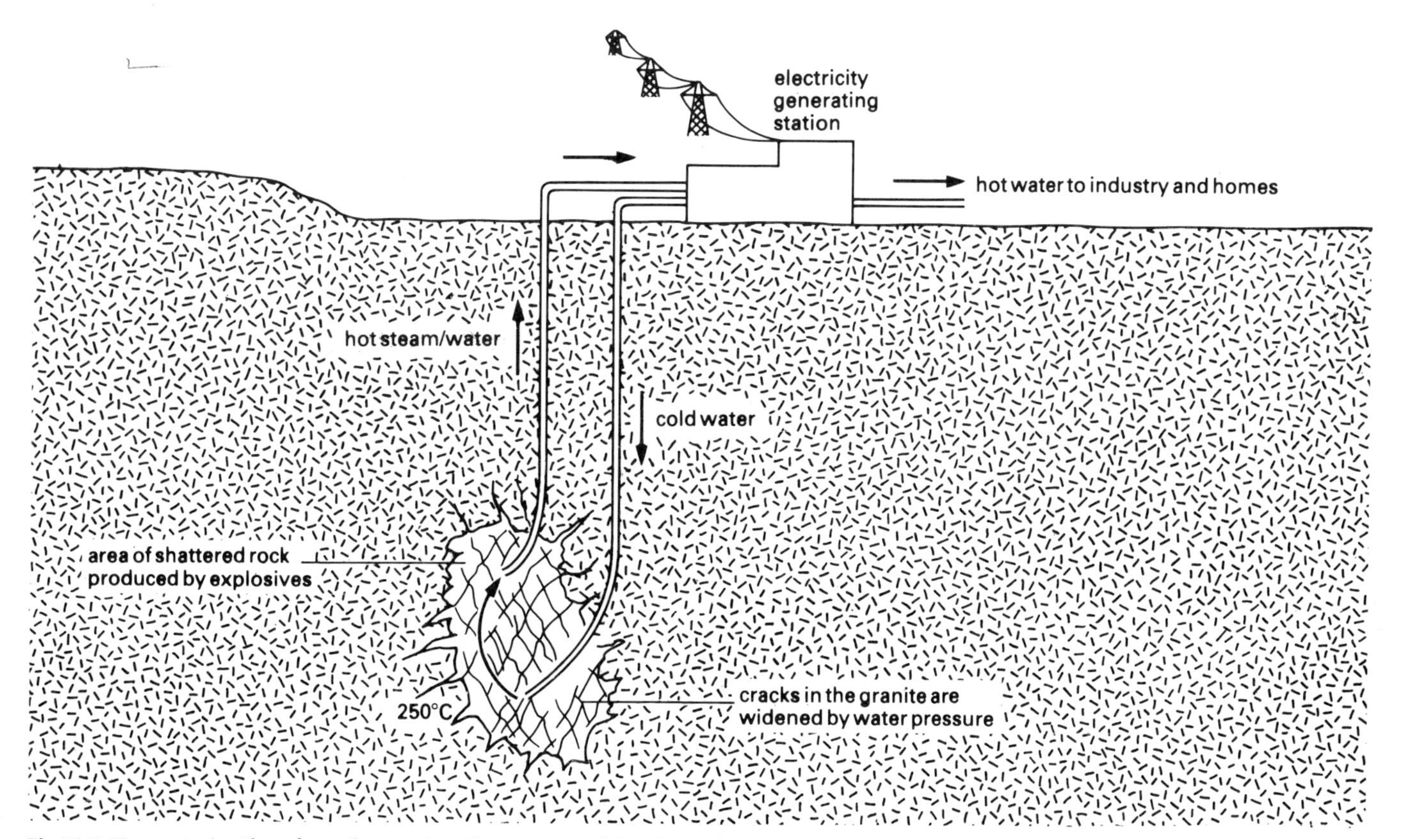

Fig 16.7 How to 'mine' heat from dry granite. The steam reaching the surface could be used to drive turbines to generate electricity.

Work has now begun on drilling boreholes down to 6000 metres. The temperature is 250 °C at this depth. This will produce superheated steam. It will be possible to drive turbines for making electricity. This method could provide an important new source of energy. Research is going on to discover other suitable sites around Britain. The site at Cornwall was selected because:

a) it is located on suitable granite.
b) there are good roads to bring in equipment easily.
c) there is a plentiful water supply.

Activity 1: How water is heated by rocks

1 Heat equal quantities of gravel and sand, in a metal dish in an oven for about one hour.
2 Place the warmed mixture in a metal eureka can, with a glass tube in position as shown (Fig 16.8).
3 Record the 'dry rock' temperature.
4 Allow cold water to trickle into the top of the glass tube.
5 Record the water temperature of each 500 cm^3 of water, as it leaves the can. Also record the 'rock temperature' at one-minute intervals.
6 Set out your results in a table:

Volume of water (500 cm^3)	Water temperature (°C)	Rock temperature (°C)
1		
2		
3		
4		

7 Make a graph of your results – temperature against volume.

Questions

a) What is the temperature *difference* between the water and the gravel-sand mixture? Does this difference remain the same?
b) How efficient is water in removing heat from the gravel-sand mixture?
c) Are the results different if the cold water is poured in quickly?
d) Explain fully what is happening in this experiment.

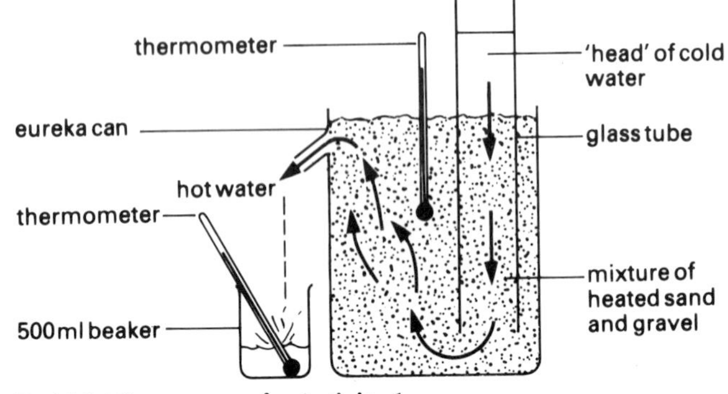

Fig 16.8 The apparatus for Activity 1

Exercise 1: Rocks and their products

There are at least 65 words hidden in the puzzle. They are printed horizontally, vertically and diagonally. They may be forwards or backwards.

L	A	T	S	Y	R	C	C	F	E	L	D	S	P	A	R	P	G
E	L	O	H	E	R	O	B	O	D	I	L	O	S	H	T	A	B
O	P	E	N	R	D	O	T	C	P	K	A	S	S	I	S	I	S
U	S	E	E	A	S	T	I	P	E	P	T	P	A	T	L	N	H
T	W	P	E	V	E	I	N	S	R	R	E	D	L	O	S	T	A
N	I	L	O	A	K	A	P	R	M	E	M	R	G	A	N	I	T
P	Y	A	L	C	M	O	L	T	E	N	A	C	I	M	C	H	T
C	R	C	I	D	I	C	A	M	A	E	T	S	O	R	E	E	E
E	R	U	Q	S	N	E	T	F	B	U	U	L	S	A	T	Z	R
R	A	M	U	D	E	S	E	A	L	K	R	A	T	A	N	E	E
A	U	N	I	A	R	L	S	A	E	C	B	M	L	O	P	T	D
M	Q	C	D	S	I	D	N	A	S	I	I	S	R	A	P	O	T
I	E	N	O	T	S	D	A	O	R	R	N	B	P	K	C	O	R
C	S	I	L	V	E	R	H	O	T	B	E	S	E	T	A	L	P
S	L	Z	X	I	M	E	X	P	L	O	S	I	V	E	S	S	U

(*Solution on page 144*)

feldspar
top
gas
paint
bath
silt
pipe
heat
tinplate
molten
permeable
steam
open
paper
rock
brass
tin
trim
brick
slate
up
tar

veins
lead
copper
metal
solder
glass
acidic
pits
clay
liquid
kaolin
placer
tap
zinc
silver
hot
ceramics
tools
seal
rain
solid
mix

mica
ore
quarry
miner
tile
turbines
explosives
shattered
use
out
plates
bronze
crystal
land
sand
roadstone
borehole
soil
pass
mud
renew

When you have found all the words, look at the unused letters. They spell out a secret message across the puzzle. Can you discover what this is?

_____ (5)

___ (3)

______ (6)

Exercise 2

1 Complete each statement with the most accurate phrase.

a) The scenery of Devon and Cornwall is made of
   A slate uplands and granite lowlands
   B granite lowlands and shale uplands
   C granite uplands and slate lowlands
   D granite uplands and shale lowlands

b) Mineral veins are formed by
   A the rocks cracking and moving apart
   B hot watery solutions in cracks
   C hot magma intruding the rocks
   D earthquakes and volcanic activity

c) A metamorphic aureole is formed by
   A intense heat and pressure
   B intense folding
   C heat loss from an intrusion
   D an intrusion pushing the rocks apart

d) Tin is used to make
   A bronze
   B brass
   C steel
   D aluminium

e) Geothermal energy is greatest in
   A granite rocks
   B active volcanic regions
   C synclinal (downfolded) limestones
   D deeply-buried shales and slates

2 Write out and complete the following passage, using the words from the list below.

The rocks between the Mendips and Bath are downfolded into a ________ structure. The limestone is ________. Water is trapped between the ________ above and the ________ rocks below. The water is directed to one site at Bath by a ________. The water rises because of ________ to produce a hot ________. It takes ________ years for the water to reach ________ from the Mendips. The temperature at the hot spring is ________. At a depth of two and a half kilometres the water is ________.

80 °C 46.5 °C 2000 syncline Bath Silurian faultline coal measures spring permeable pressure

3 Correctly pair the phrases together to form proper sentences:

a) The large intrusions are called / granite and slate.
b) Different metallic ores are found / because of erosion.
c) Cornish tin is mined / using high pressure hoses.
d) The white kaolin spoil heaps could be used / to make plate glass.
e) In Cornwall, traditional local building materials are / bathyliths.
f) The granite is now exposed as moorland / in mineral veins.
g) Kaolin is washed out of the rocks / deep below the surface.
h) Molten tin is used / in the building industry.

Exercise 3

1 Discuss how the economy and environment of Devon and Cornwall have been affected by the local rocks.

2 Explain the formation of granite and slate.

3 Explain fully how mineral veins are formed.

4 Draw a diagram to show the zones of formation of tin, copper, lead and zinc. Use symbols to show the distance of each zone from the granite intrusion. Use Fig 16.2 as a guide.

5 Write a description of tin mining, and its economic importance.

6 a) Explain the formation of kaolin (china clay).
  b) How is it recovered?

7 Construct a pie graph to show the main uses of kaolin. Use the figures shown in Table 2.

8 Explain why the quarrying of kaolin has had a damaging effect on the environment. Are there any remedies?

9 a) What is *geothermal energy*?
  b) Where was geothermal energy first used in Britain?

10 Copy Fig 16.6 and fully explain why hot water reaches Bath under pressure.

11 a) Why are granites hotter than most other rocks at depth?
   b) Explain fully how this heat can be extracted and used.

17 FUEL RESOURCES

Fig 17.1 Fossil fern leaves. Plants growing in lowland coastal swamps were later submerged to form coal. Plant fossils, such as these fern leaves, are often found in coal.

Coal, oil (petroleum) and natural gas, are important *fuels*. A fuel is burned to produce *heat energy*. This heat can be used directly for furnaces and home heating, or indirectly to produce electricity, and engine movement. In Britain 95% of our energy is produced by three sources – coal, oil and natural gas. These are known as *fossil fuels*, because they formed from the remains of plants and animals that were buried underground for millions of years (Fig 17.1).

RENEWABLE AND NON-RENEWABLE RESOURCES

There are two kinds of energy resource: (a) *renewable* energy that can be replaced; (b) *non-renewable* energy cannot be replaced. Water power is renewable. It is always available as long as there is rain. The fossil fuels are examples of non-renewable resources. Once burnt they are gone forever.

By the year 2000, oil will have become very scarce. Britain's oilfields are now at their peak production. They will continue to decline from now on. On the other hand, coal is likely to be increasingly used in the future. It will supply part of our energy needs well into the twenty-first century and beyond.

COAL

Three hundred million years ago, in the Carboniferous Period, large areas of Britain were covered by swampy forests (p.116). The plant debris from the forests steadily collected in the stagnant waters. Most of the dead material did not rot away because the acidic swamp waters prevented the decomposition. Instead, fibres from the roots and stems of plants matted together to form *peat*. This was the first stage in coal formation.

The change into coal

The layers of peat later became buried and compressed under thousands of metres of sediment (Fig 17.2). Any water, hydrogen or oxygen present in the peat was squeezed out by the pressure. This process left behind a high concentration of black carbon in the form of coal.

A layer of underground coal is called a *seam*. It takes six metres of peat to form a coal seam one metre thick. Many coal seams are no more than a few millimetres thick. Most of the workable seams in British coalfields are one to two metres in thickness. Coal seams may be sandwiched between layers of sedimentary rocks. Underneath many coal seams is a layer of *seat earth*. This was the forest soil. It is often found with fossils of tree roots preserved in it.

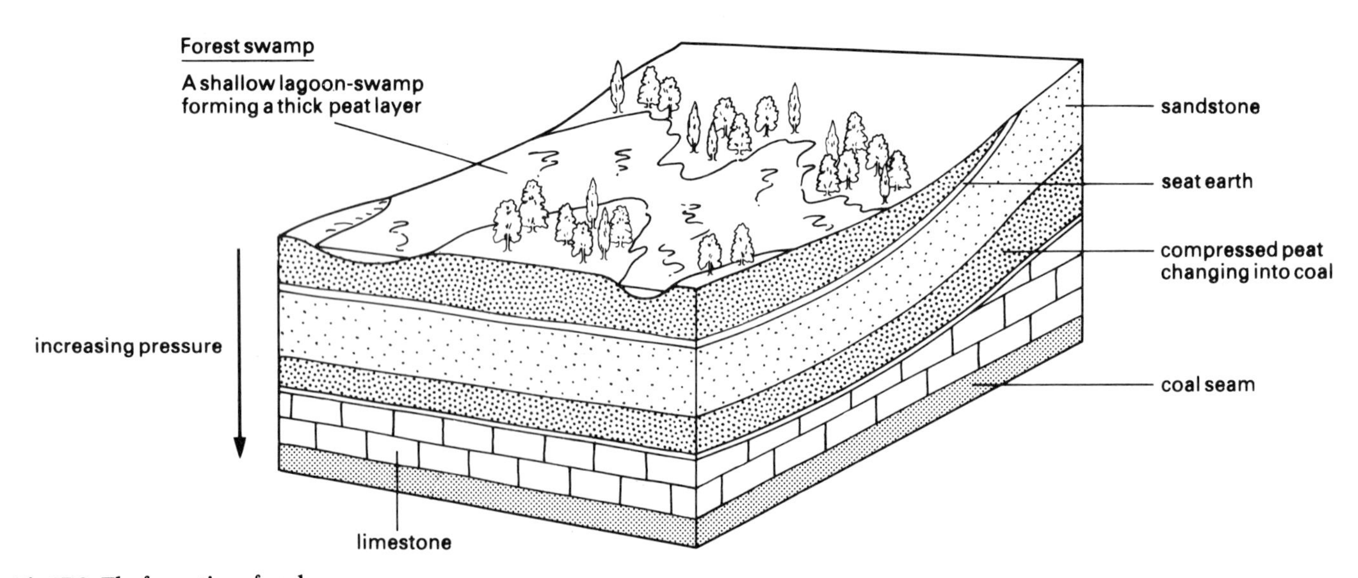

Fig 17.2 The formation of coal

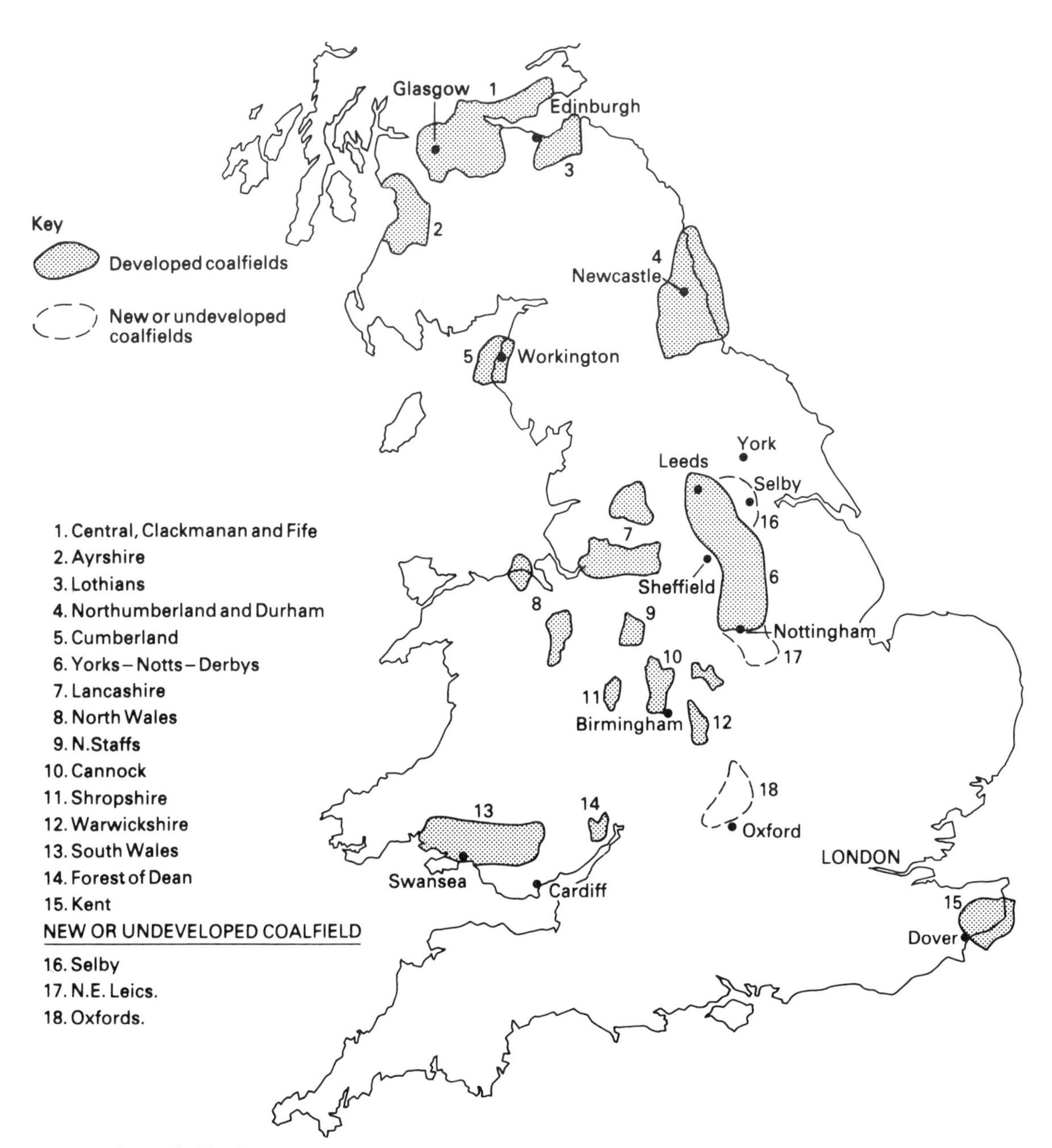

Fig 17.3 The coalfields of Britain

Coalfields

Underground seams of coal, and the layered rocks sandwiched between them, make up a *coalfield* (Fig 17.3). The uplift of the Pennine hills has produced two major coalfields on either side of the *anticline* (upfold) (Fig 17.4). The crest of the anticline, together with its coal, has long since been worn away by the action of ice and water.

Coal extraction

There are three main methods of extracting coal: (a) opencast, (b) deep mining and (c) drift mining.

a) Opencast mining

The National Coal Board has 63 opencast sites. They produce over 14 million tonnes per year (1983); 12% of total UK coal production. First the overlying soil and rock is removed. This is known as the *overburden*. The exposed coal seam is then broken up and removed by a huge walking *dragline* (Fig 17.5).

Advantages:

(i) If the seam is near to the surface, it is a much quicker method of extraction than underground mining. Fewer men are needed to produce each tonne of coal.

(ii) There are no expensive tunnels with underground props, lifts, water pumps, and other machinery, to build and maintain.

(iii) There are no underground workings. This cuts down the risk to human life, caused by rockfalls, explosions and flooding.

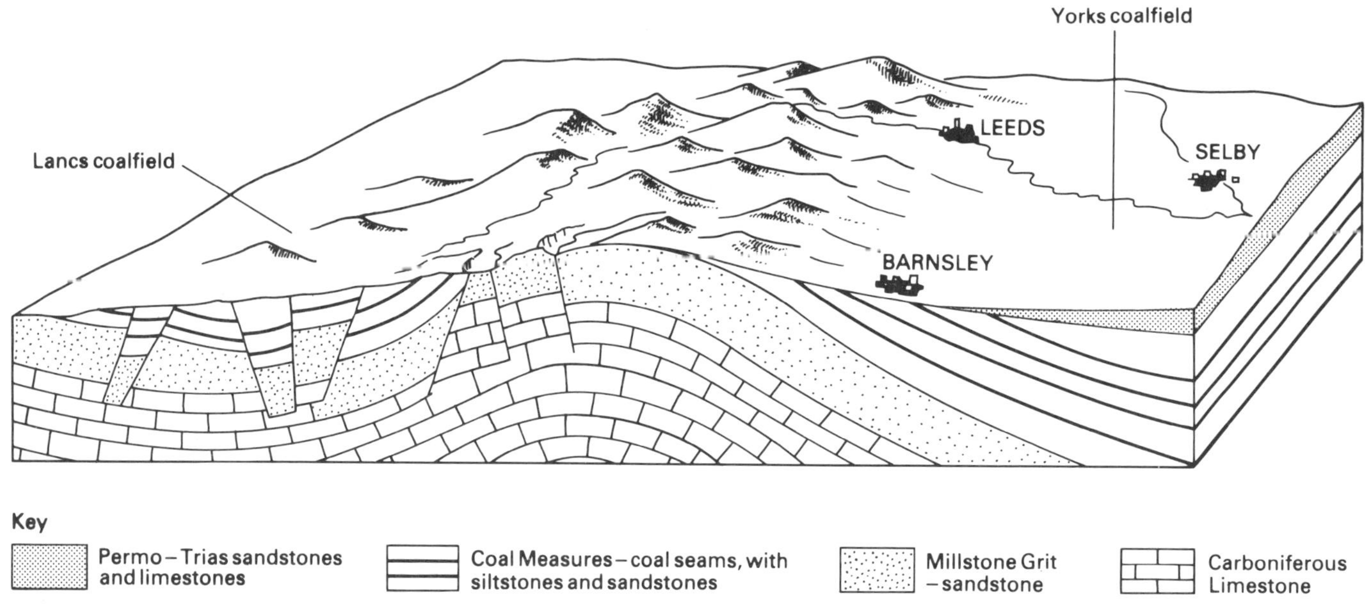

Fig 17.4 The Pennine anticline and its coalfields

Fig 17.5 Opencast mining

b) Deep mining

Deep-mined coal is reached by roadway tunnels that lead away from a vertical shaft. Nearly all the coal is cut away by *shearers*, machines with giant rotating blades (Fig 17.6). The shearer moves along the seam face for about 170 metres dropping the coal it cuts onto a conveyor belt beside it. The belt carries the coal back to waiting wagons. An engine hauls the wagons to the coal shaft and they are lifted to the surface. This method is called *longwall* mining.

Once the coal face has been cut by the shearer, the line of powered roof supports moves forward to hold up the new roof. Power supports are like enormous car jacks. The area behind the supports is then allowed to collapse and cave in.

c) Drift mining

Europe's biggest and newest drift-mining operation is at Selby in North Yorkshire. This is a huge new coalfield about the size of the Isle of Wight; 110 square miles in area (Fig 17.7). Its total reserve of coal amounts to 2000 million tonnes and is expected to last for about 150 years, depending on the rate of extraction. The Barnsley seam beneath the Selby area, is three and a half metres thick in some places – the thickest of any British coal seam. The National Coal Board have spent £1000 million developing the five mines at the Selby complex. Two shafts have been sunk at each mine to take men and materials underground. They also provide ventilation for all the interconnected workings. Each one of the five mines is working four coal faces at a time. Production of coal began at the Wistow mine on 4 July 1983. By 1988 an

Fig 17.6 A coal shearer moving along the coal face

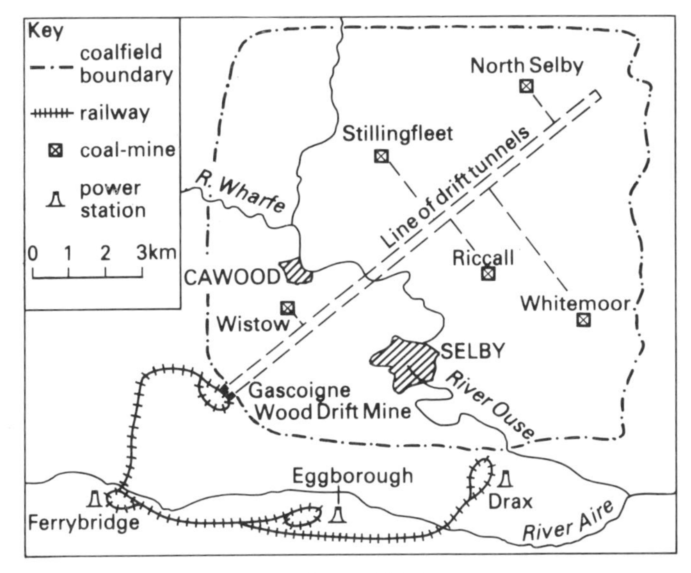

Fig 17.7 Sketch map of the Selby coalfield

overall production of 10 million tonnes a year is expected.

How the Selby coalfield was developed
The first step was to build two parallel tunnels called *drifts*. These tunnels slope downwards from the surface at Gascoigne Wood, and each one is 15 kilometres long. These tunnels make up the central transport system for the complex (Fig 17.8). They house conveyor belts that carry coal from the five mines to the surface.

The coal-carrying conveyors work 24 hours a day and are much more efficient than the stop-go method of winding coal wagons up a vertical shaft. Also, the movement of coal is now easier to monitor and control, by computers, from the surface.

The conveyor belts can handle up to 3000 tonnes of coal an hour. At the surface, computers control the washing and grading of the coal. It is loaded automatically onto trains and transported to the Drax, Eggborough and Ferrybridge power stations.

The environmental problems
In the past, the extraction of coal in large quantities meant the production of large heaps of waste. Many of the early waste tips have been landscaped and are even being used as farmland. New mine workings, like Selby, are carefully landscaped to blend into the countryside as much as possible.

The removal of a three-metre thick layer of coal, as in the Barnsley seam, will inevitably mean the collapse of the overlying rocks once a section has been worked. Subsidence can cause a lot of structural damage as the land surface resettles. Drains may crack open, and roads and houses may be damaged.

Mineral oil (petroleum)

Petroleum, somctimes called *crude oil*, is a dark brown, sticky mixture. It is made of chemicals called *hydrocarbons* (hydrogen and carbon compounds). Petroleum forms slowly in the sea from organic

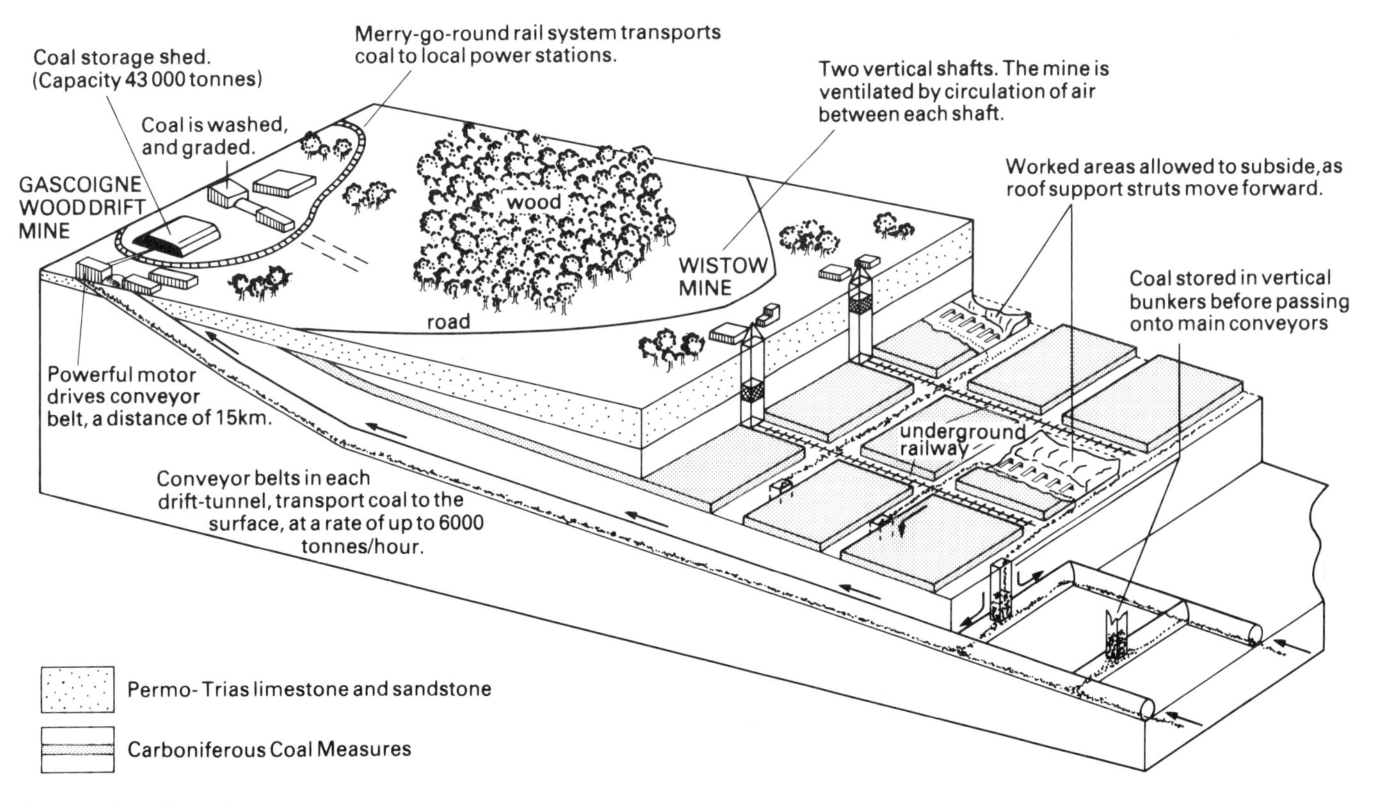

Fig 17.8 The Selby drift mine system

material which collects on the sea bed. It mixes with muddy sediment in stagnant conditions. Bacteria bring about partial breakdown and decay as they use up the oxygen in the rotting matter. Hydrocarbons are left behind as an organic sludge. This eventually becomes part of the muddy sediment on the sea floor, and in time, the layers of mud become buried by new deposits. The mud is put under pressure by the weight of overlying sediment and becomes compacted. Eventually the various hydrocarbon compounds combine to form *petroleum*. The formation of oil in this way takes about 19 million years. This is faster than the formation of coal, which takes over 100 million years.

The fine-grained shale (mudstone) in which petroleum first forms is known as the *source rock*. In many cases, pressure eventually squeezes most of the petroleum out of the source rock. The tiny droplets of oil collect together and move upwards. The lighter hydrocarbons form natural gas. Both oil and gas may collect in the pore spaces of rocks such as sandstone (Fig 17.9). A porous rock that holds oil or natural gas in this way is called a *reservoir rock*. Oil and gas can continue to migrate through porous rocks, until they reach the surface.

Oil traps

The migration of oil, and its loss at the surface, may be stopped by impermeable rock layers. An impermeable rock does not have pore spaces large enough to allow the oil to migrate in any quantity. It acts as a *caprock* eg shale, mudstone and salt. Rock structures that make oil and gas concentrate underground are known as *traps*. About 80–90 per cent of known petroleum reserves are found in dome-shaped upfolds called

(a)

(b)

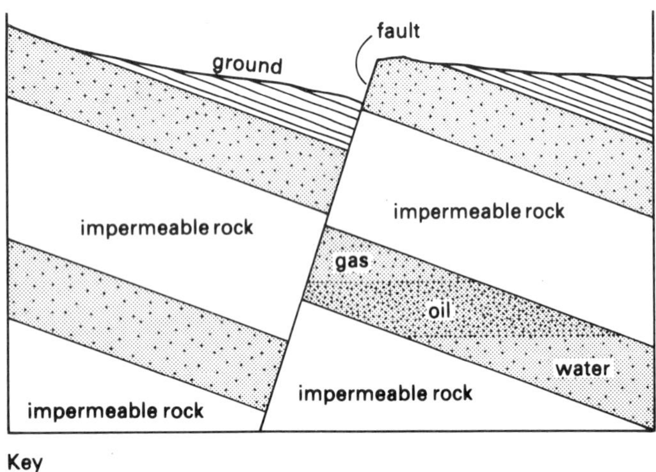

Fig 17.9 Oil traps (a) an anticline (b) a fault

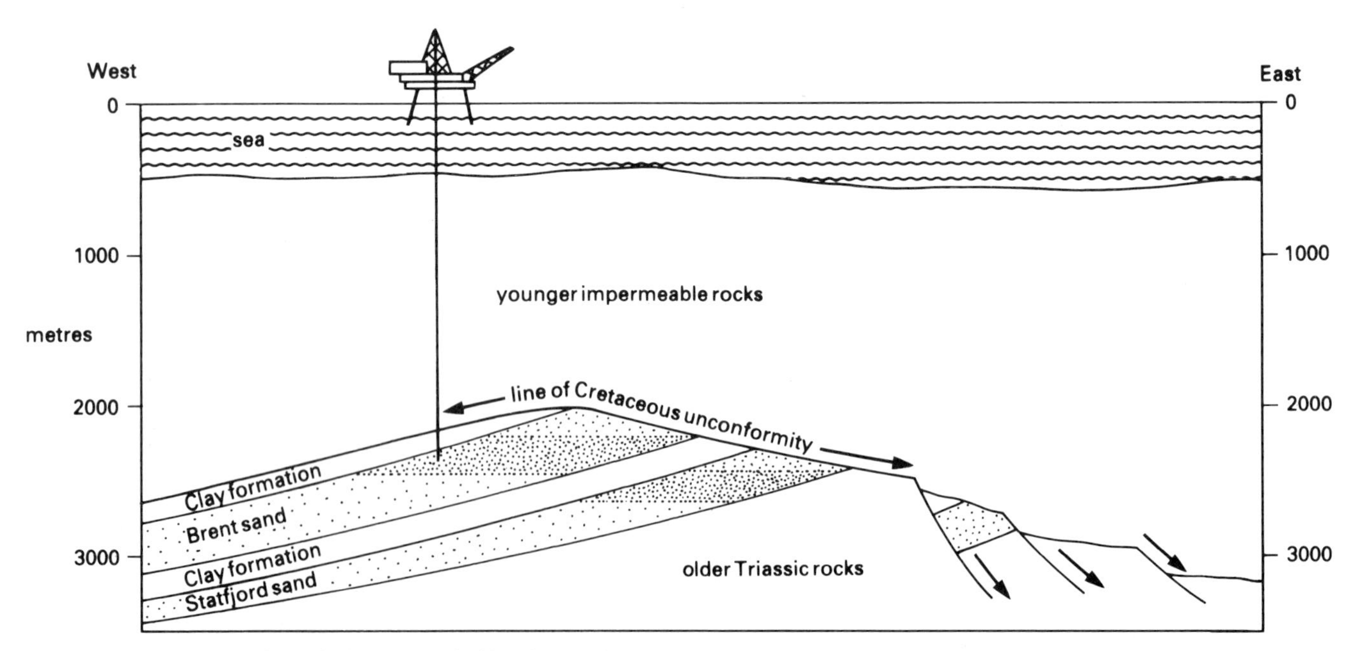

Fig 17.10 Cross-section through the Brent oilfield in the North Sea. The permeable sandstones are Jurassic in age. Oil is prevented from escaping by the impermeable rocks above the unconformity.

anticlines and in fault-traps.

In an anticlinal trap (17.9a) the oil migrates into the area of least pressure in the fold crest. The lighter hydrocarbon gases are often found at the top of the fold crest. The largest known oilfield of this type in the world, is the Ghawar field in Saudi Arabia. It is 240 km long and 35 km wide – about the same area as Devon and Cornwall put together!

The slipping movement of rocks along a fault plane can bring an impermeable rock layer against the porous reservoir rock and seal it (17.9b). This prevents the oil from escaping.

One of the largest oilfields in the North Sea is the Brent field (Fig 17.10). Here the oil reservoir is formed in sandstones of Jurassic age. The caprock is impermeable Cretaceous rock that lies at a different angle. This forms a structural trap known as an *unconformity*.

North Sea oil and gas

Britain is now self-sufficient in oil and natural gas. This is due to the large number of productive fields that have been found under the North Sea (Fig 17.11). Although most of these fields are small, they are fairly numerous.

The North Sea is a good area for oil and gas production for several reasons.

a) The region has been a sedimentary basin for the last 400 million years, so there are plenty of source rocks.
b) There are large thicknesses of permeable and impermeable sediments.
c) The area has undergone gentle, slow folding and faulting; this has formed ideal oil-trapping structures.

Activity 1: To show how oil migrates from one rock to another

The permeable rocks in this model are represented by cellulose sponges.

1 Place a cellulose sponge in a bowl and soak it with water. This represents the source rock containing oil.
2 Place a dry cellulose sponge on top of the wet sponge, followed by a layer of impermeable material, such as a piece of rigid plastic.
3 Finally, place a third cellulose sponge on top.
4 Examine the layers after a few minutes. What happened to the second sponge? What happened to the third sponge? Explain your observations.
5 Repeat the activity, but this time use two halves of plastic sheet. Apply increasing pressure with your hand, to the topmost sponge. Note what happens.

Questions

What effect does overlying pressure have on the rate of upward movement? How is the topmost sponge affected?

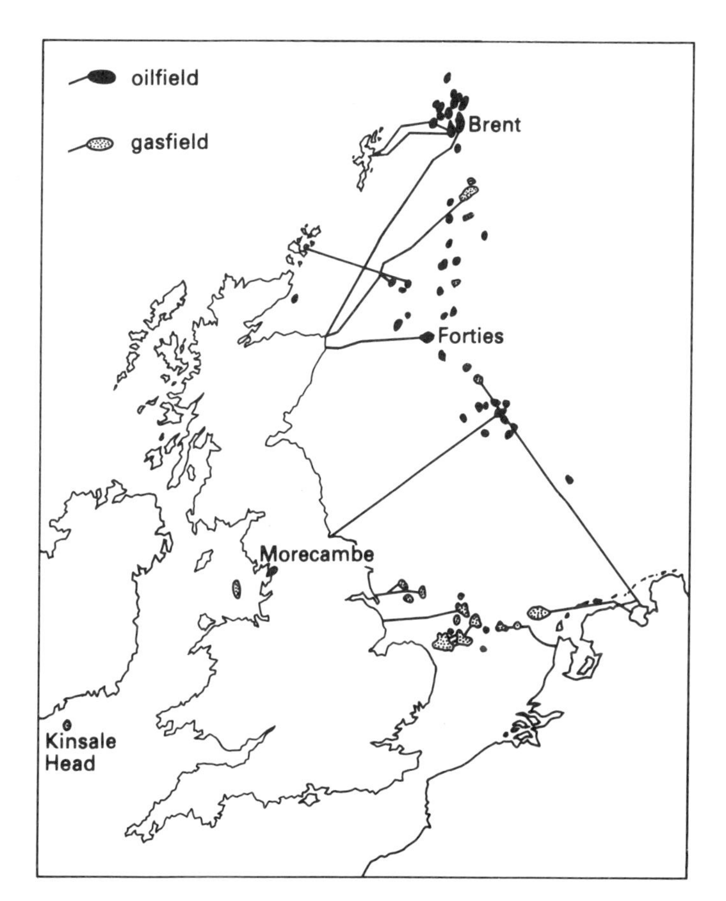

Fig 17.11 The oil and gas fields around Britain

Oil exploration

The first step in looking for oil below the North Sea is to hunt for likely rock structures that might form oil traps. This is mainly done by *seismic surveying* (Fig 17.12). The survey ship tows a compressed air 'shot' behind it together with a line of receptors called *geophones*. At regular intervals compressed air 'pulses' are released to make shock waves. They pass into the rocks below the sea bed, and bounce back off each rock layer. The 'echoes' are picked up by the geophones and recorded for processing by computer. A picture of what the rocks look like underground is built up. In this way it is possible to find likely oil-trap structures. The only way to be certain they contain oil is to drill a borehole. What section of the reflecting rock layer in Fig 17.12 might contain oil?

The borehole is cut by a rotating diamond studded *bit* lowered from the rig. Lubrication mud is pumped down the hollow drill pipe to cool the bit and wash away rock chippings. The mud returns to the surface through a pipe around the outside of the drill pipe. The mud is kept under pressure to prevent a *blow out*. This is a sudden explosive surge of gas or oil, caused by the drill hitting an underground pocket of oil or gas under pressure. Drilling rigs are built to provide floating exploration platforms that can be moved from one oilfield to another. Their large size and complex design

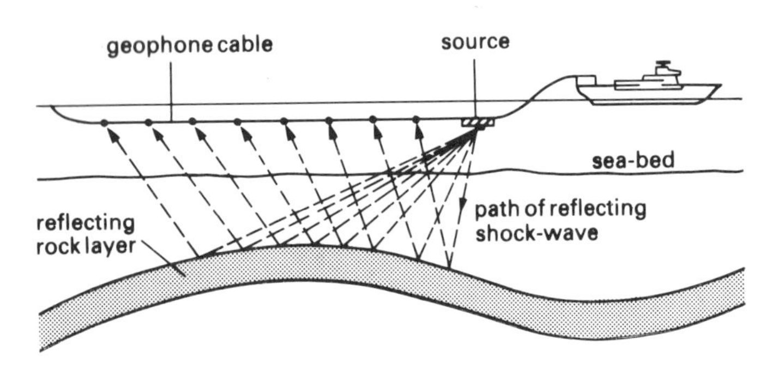

Fig 17.12 A seismic survey of rock layers below the sea bed

means that they are very costly to build, and are worth over £100 million each.

Once the well has been drilled the exploration rig moves on. If oil has been found, the rig is replaced by the *production platform*. Unlike the rigs, these are usually fixed to the sea floor. They are massive structures built out of steel and concrete. The platforms weigh over 300 000 tonnes and cost over £150 million each to build and stand 230 metres high from the sea bed. By comparison Big Ben is 97 m high.

Apart from its great height, the production platform must have a large enough area to house the crew accommodation, helicopter deck, remote control facilities and drilling equipment.

The importance of North Sea oil
Only about four per cent of the world's total oil supply comes from the North Sea, but this is more than enough to supply the needs of the UK. The UK is now the fifth-largest oil producer in the world, with total reserves in the North Sea estimated at between 2000–4000 million tonnes. Yearly output could decline after 1985 unless many new fields are found. By 1990 it is possible that the UK may be a net importer of oil again.

THE USES OF FUELS

Coal, oil and natural gas are mainly used to produce heat and electricity. The physical character of each fuel affects how it is transported and used. Coal is a solid, bulky material. It is more difficult to transport than oil or natural gas. It is therefore an ideal fuel for supply in bulk to one location, such as a power station or steelworks, near a coalfield. By comparison, oil and natural gas are easier to transport. For instance, gas is carried direct to consumers by pipeline straight from under the North Sea.

Oil is refined to produce high energy transport fuels like petrol and aviation fuels. The transport fuels can easily be stored in the fuel tanks of vehicles. Oil is also converted to a wide range of *petrochemicals* that are extensively used in industry. As oil becomes scarcer it

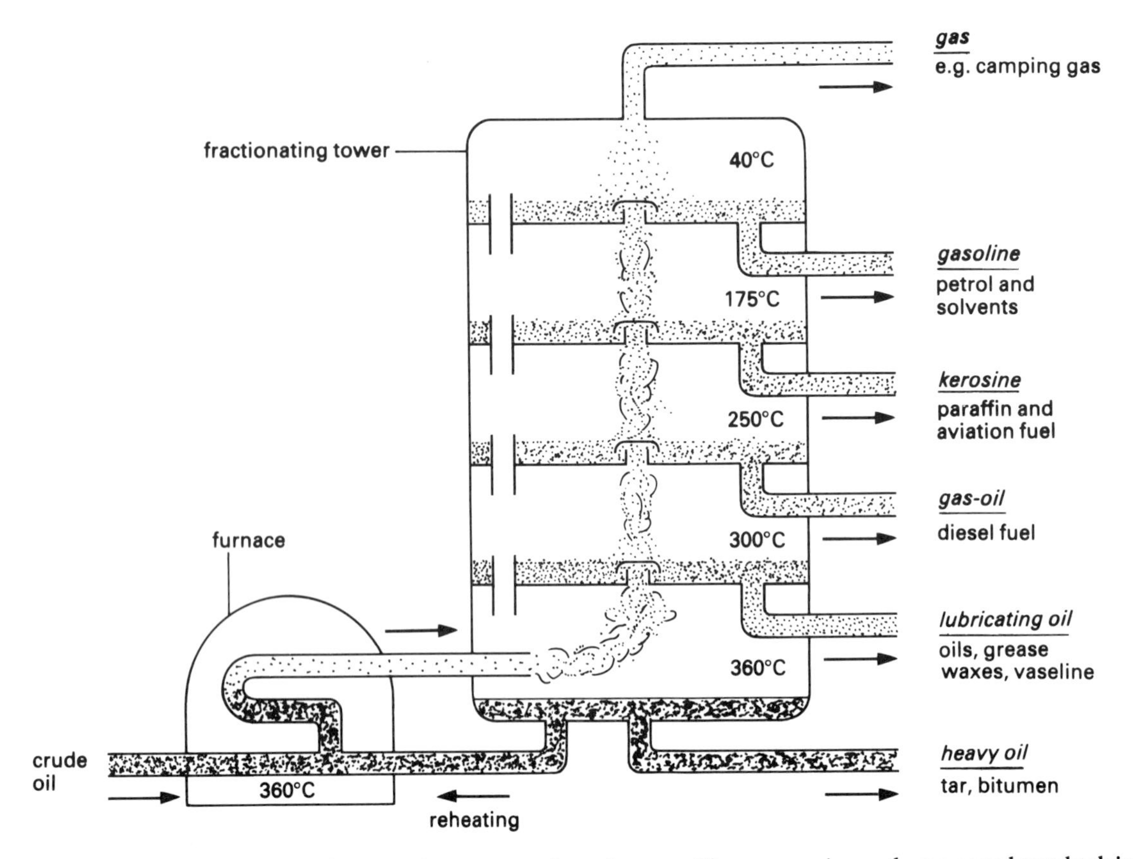

Fig 17.13 The crude oil enters the fractionating tower as a heated vapour. The vapours rise up the tower and turn back into liquids at various levels. Liquids with high boiling points collect near the bottom. Those with low boiling points condense near the top.

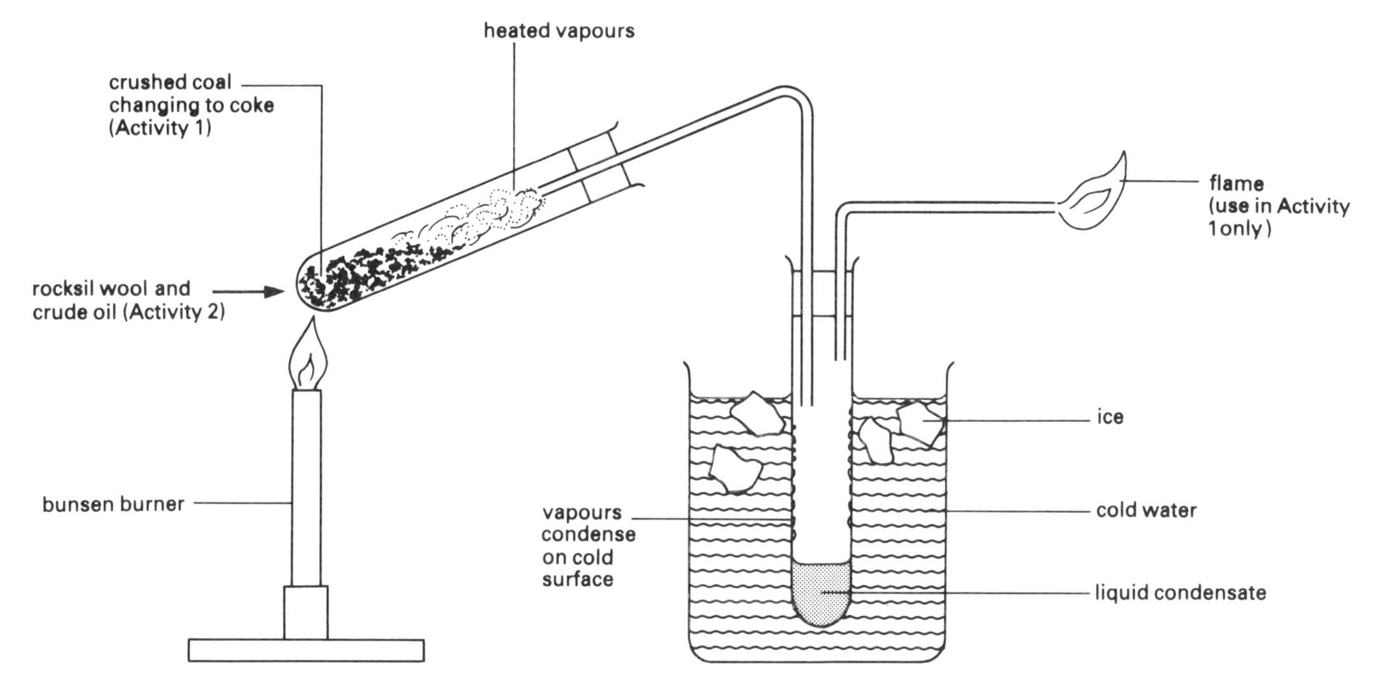

Fig 17.14 Apparatus for Activities 1 and 2

would be better to use it for these purposes, rather than as a fuel in power stations.

Table 1 Fossil fuel usage in the UK (1981)

	%
Making of electricity	30
Heating in schools, homes and offices	27
Industry	21
Transport	17
Petrochemicals	5
	100

PRODUCTS OF FOSSIL FUELS

Many everyday items are made out of petrochemicals. Clothes, cosmetics, furniture, car parts, plastic containers, paints, floor tiles, video-tapes, fertilisers, insecticides, eyeshadow, skin cleanser, nail lacquer, lipstick, shampoo – the list is almost endless. Most of these products are made from the petrochemicals that come from the refining and *distillation* (breaking down) of coal and petroleum. About 90% of the world's organic chemicals come from petroleum, mainly because it is easier to transport and process compared to coal.

Distillation of coal

If coal is heated in a *vacuum* (without air) it will break down into *coke, ammonia, tar* and *coal-gas*. Solid coke is a useful by-product and is used as a smokeless fuel in furnaces. Tar and ammonia are processed further to give a wide range of chemical products. It is even possible to make fuel oils from coal but the process is very expensive. No doubt coal will be increasingly used as a source of petrochemicals when oil begins to run out.

Distillation of oil

Fractional distillation of crude oil is the first step in the process of refining. This takes place in a *fractionating tower* 45 m high. The hydrocarbons in petroleum (crude oil), are separated into *fractions* by heating to 360 °C (Fig 17.13).

Activity 1: Distillation of coal

It is possible to reduce solid coal into liquids and gas, using the apparatus shown in Fig 17.14. Solid coke, liquid tar and ammonia, and coal-gas can be produced by this method.

1. Crush some coal to a powder and place it in a test tube. Strongly heat the powdered coal.
2. Light the gas which comes out of the pipe.
3. Watch how a liquid forms and collects in the bottom of the test tube in the water-filled beaker.
4. When the gas ceases to burn, turn off the heat, and allow the apparatus to cool.
5. Pour some of the condensed liquid onto a watch-glass. There is a black substance, together with a clear liquid. One substance is ammonia; the other is tar. Can you tell which is which?
6. Finally, examine the coke that remains, and describe what it looks like.

Activity 2: Distillation of crude oil

Compare the results of this activity, with the distilled fraction produced in a fractionating tower, as shown in Fig 17.14.

1 Soak rocksil wool in 5 cm^3 crude oil and place in the apparatus as shown in Fig 17.14. (You do not need the flame tube in this activity.)
2 Heat the oil-soaked wool gently.
3 Collect approximately 1 cm^3 of the watery-looking liquid that distils in the test tube. This is the first low temperature fraction.
4 In a second test tube collect another 1 cm^3 of distilled liquid, and increase the heating as necessary.
5 Repeat this procedure to produce three more fractions. As heating is increased, the liquid that distils becomes thicker and darker.
6 Pour each fraction onto a separated numbered watch-glass. Make a note of the appearance and smell of each one.
7 Soak a piece of cotton-wool in the liquid on each watch-glass. Carefully set fire to each piece in turn.
   a) Which fraction burns the brightest and longest?
   b) Which fraction do you think would make the best quality fuel?
   c) Using Fig 17.13 as a guide, what names would you give to the five fractions produced in this activity?

Exercise 1: The oil-strike game

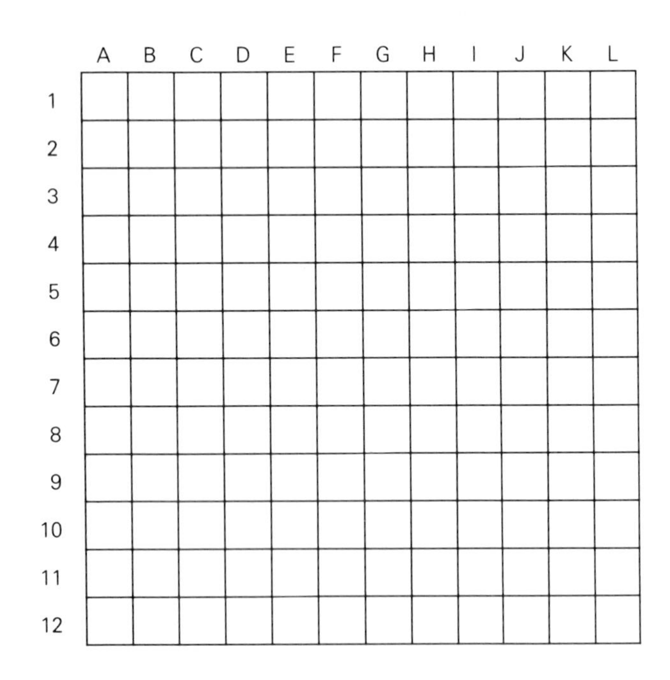

Key
3 fault traps = 2 squares
2 unconformity traps = 4 squares
1 anticline trap = 9 squares

The shape of the traps does not have to be the same. For example:

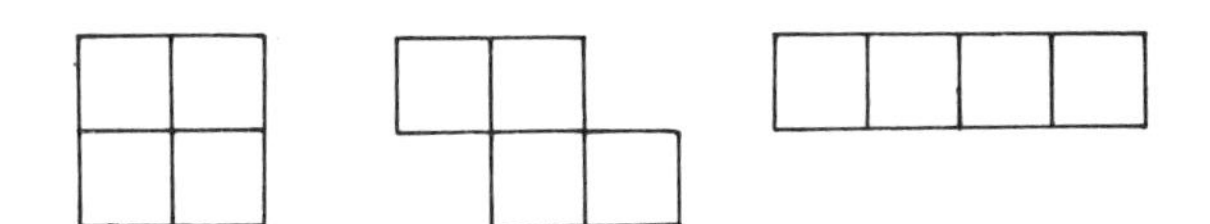

Rules

1 Play in pairs. Each player makes a copy of the grid.
2 Secretly position oilfields on the grid in one colour. You are allowed 3 fault traps; 2 unconformity traps and 1 anticline trap.
3 Take turns at 'drilling' for your oilfields in your opponent's area. Do this by calling out the reference for each square eg A1, B2,and so on.
4 Record all the 'strikes' and 'misses' of your opponent in a different colour. An oilfield is in full production when there is an oil well on each of its squares. When this happens you are allowed an extra turn.
5 During the game each player is allowed one *seismic survey*. This is done by asking if there are any oilfields along one line, eg, line F or line 8. The opponent must answer 'Yes' or 'No'.
6 When you have all the oilfields on your opponents grid in full production, you have won the game.
7 It costs £1 million to drill each oil well but each productive well makes £5 million profit.
8 Count up the number of 'dry' wells and 'strike' wells. Did you make a profit?

Exercise 2

1 Why are coal and oil called 'fossil' fuels?
2 What is the difference between a renewable and non-renewable resource? Give examples of both kinds.
3 What is the *first* stage in the formation of coal?
4 What is an underground layer of coal called?
5 What is the thickness of peat needed to form a layer of coal 1 metre thick?
6 What is a 'seat earth'?
7 Study Fig 17.4
   a) Make a copy of the front section-view of the diagram and label:
      the crest of the anticline;
      the faults;
      the coal seams.
   b) Which of the two coalfields will be easier to work? Why is this?
   c) Why are coal seams missing from the top of the Pennines?
8 The following terms are all to do with mining coal. Explain what each one means:
   overburden dragline shearers longwall-mining powered roof-supports drift tunnels subsidence
9 The following terms are all to do with oil. Write a sentence explaining each one:
   hydrocarbons oil-shale source-rock reservoir rock oil-traps

10 Complete the following passage using words from the list below:

Petroleum moves through ________ rocks. It becomes concentrated beneath ________ rocks. Anticlines, ________, and unconformities are the main kinds of ________. Rock structures where oil is likely to be trapped can be found by ________ surveying. Shock ________ pass into the rocks below the sea ________. The 'echoes' are picked up by ________ and recorded. This way it is possible to ________ where the best structures are. Not all of the structures will contain ________. The only way to find out is to ________ a borehole.

impermeable seismic bed geophones waves
permeable oil-trap map petroleum drill faults

11 Why is coal well suited for use as a fuel by power stations?

12 Why is oil an ideal transport fuel?

13 a) Make a pie graph to show the uses made of fossil fuels (see Table 1).
b) What are the main uses of petrochemicals?

14 Look carefully at Fig 17.13. What happens in a fractionating tower?

Exercise 3

1 a) What is meant by a 'fuel'?
b) To what extent is Britain reliant on fossil fuels, and how might this change over the next 15 years?

2 Explain the formation of peat and how it is changed to coal.

3 a) Write a brief description of open-cast mining.
b) What are the main reasons for open-cast mining being so efficient?
c) Why does open-cast mining only account for 12% of total coal production?

4 What are the advantages of drift mining, compared to conventional deep mining?

5 Describe two problems created by the large-scale mining of coal.

6 a) Explain how oil forms.
b) Why does it migrate and collect in permeable rocks?

7 a) What is an oil-trap?
b) Explain in notes and sketches the three main kinds of oil-trap. Label the reservoir rocks on each of your sketches.

8 a) Where are most of the British oil and gas fields?
b) Outline the main reasons for the formation of these fields.

9 Use graph paper to make a scaled diagram showing the relative heights of the following:
an oil production platform
Big Ben
your school building.

Appendix 1: MINERAL IDENTIFICATION

THE COMPUTER PROGRAM

The program is written in BASIC for a Sinclair Spectrum microcomputer, but is easily adaptable to other machines. It is based on the flow diagram shown in Fig 7.14. Although the program is not handling very complex information, the application of a computer to mineral studies in this way makes identification and testing a more interesting process. The program structure is as follows.

Program

```
 1Ø REM CSE MINERAL IDENTIFICATION
 2Ø PRINT "CHOOSE A SPECIMEN"
 3Ø PRINT
 4Ø PRINT "ANSWER QUESTIONS Y OR N"
 5Ø PRINT
 55 LET A = Ø
 6Ø FOR N = 1 TO 20
 7Ø READ O$: READ C$
 8Ø PRINT: PRINT "HAS THE SPECIMEN"; O$
 9Ø INPUT A$
1ØØ IF A$ = "N" THEN GO TO 15Ø
11Ø IF A$ = "Y" THEN GO TO 13Ø
12Ø GO TO 9Ø
13Ø PRINT: PRINT "THE MINERAL IS"; C$
14Ø GO TO 18Ø
15Ø NEXT N
155 LET A = A + 1: IF A = 3 THEN GO TO 44Ø
16Ø PRINT: PRINT "LET'S TRY AGAIN"
17Ø RESTORE GO TO 6Ø
18Ø PRINT
19Ø PRINT "HAVE YOU ANOTHER SPECIMEN?"
2ØØ INPUT A$
21Ø IF A$ = "N" THEN STOP
22Ø IF A$ = "Y" THEN CLS: RESTORE: GO TO 4Ø
23Ø GO TO 2ØØ
24Ø DATA "A SALTY TASTE?" "HALITE"
25Ø DATA "A WHITE STREAK: VIGOROUS REACTION
    TO ACID?" "CALCITE"
26Ø DATA "A YELLOW STREAK: SLIGHT REACTION
    TO ACID?" "DOLOMITE"
27Ø DATA "A BAD EGG SMELL WITH ACID?"
    "GALENA"
28Ø DATA "EASILY CRUMBLED TO A WHITE
    POWDER?" "GYPSUM"
29Ø DATA "EASILY MARKED PAPER WITH A BLACK
    STREAK?" "GRAPHITE"
3ØØ DATA "A PALE GREEN STREAK?"
    "MALACHITE"
31Ø DATA "A RED STREAK, METALLIC?"
    "HAEMATITE"
32Ø DATA "A BLACK STREAK, MAGNETIC?"
    "MAGNETITE"
33Ø DATA "A WHITE STREAK, FEELS HEAVY?"
    "BARYTES"
34Ø DATA "A THIN FLAKY CLEAVAGE; LIGHT
    COLOUR" "MUSCOVITE"
35Ø DATA "A THIN FLAKY CLEAVAGE; DARK
    COLOUR" "BIOTITE"
36Ø DATA "GREEN TO BLACK; TWO GOOD SETS OF
    CLEAVAGE AT 120° "HORNBLENDE"
37Ø DATA "GREEN TO BLACK; TWO GOOD SETS OF
    CLEAVAGE AT 90° "AUGITE"
38Ø "POOR CLEAVAGE; OLIVE GREEN?" "OLIVINE"
39Ø DATA "CUBIC CRYSTALS; OCTAHEDRAL
    CLEAVAGE?" "FLUORITE"
4ØØ DATA "CLEAVAGE AT 90°; PINK OR
    WHITE?" "ORTHOCLASE"
41Ø DATA "A GOLDEN AND METALLIC
    APPEARANCE?" "PYRITE"
42Ø DATA "A GLASSY LUSTRE: HARDNESS OF 7?"
    "QUARTZ"
43Ø DATA "A PLASTIC LUSTRE; FEELS HEAVY?"
    "SPHALERITE (ZINCBLENDE)"
44Ø PRINT
45Ø "TRY ASKING YOUR TEACHER!!!"
46Ø GO to 18Ø
```

Appendix 2: LIST OF SUPPLIERS

MINERAL INVESTIGATIONS

The specimens and specialist equipment may be obtained from any of the suppliers listed below:

Specimens
The 20 minerals recommended for study:

Silicate minerals
Quartz; orthoclase feldspar; muscovite mica; biotite mica; olivine; hornblende; augite.

Non-silicate minerals
Graphite; haematite; magnetite; galena; pyrite; sphalerite; calcite; dolomite; malachite; fluorspar; halite; barytes; gypsum.

Specialist equipment
One set Mohs scale minerals (without diamond)
Porcelain streak plate, or unglazed bathroom tile
Hand lens × 10
Geological hammer
Safety glasses

Other equipment
Steel file – small warding files are convenient
5 cm × 5 cm plate glass square
Steel penknife blade
Copper coin
Circular self-adhesive labels
Measuring cylinder
Top pan balance
Dilute hydrochloric acid
Fine paint brush
Micro-computer

GEOLOGICAL SUPPLIERS

Catalogues are obtainable on request.

Gregory, Bottley and Lloyd
8–12 Rickett Street
London
SW6 1RU

Richard Tayler Minerals
Byways
Burstead Close
Cobham
Surrey

South West Minerals Resources
14 Elphinstone Road
Peverell
Plymouth
PL2 3QQ

A number of organisations provide some free charts and literature, and 16 mm films on free loan covering many aspects of Earth Science. For catalogues of materials and films available, write to:

Association for Science Education,
College Lane,
Hatfield, Herts AL10 9AA

BP Educational Service,
Britannic House, Moor Lane,
London EC2Y 9BU

British Gas Education Service
Rm 707A, 326 High Holborn,
London WC1V 7PT

British Gas Film Library,
Park Hall Road Trading Estate,
London SE21 8EL

Central Electricity Generating Board,
Press & Publicity Office,
Sudbury House, 15 Newgate Street,
London EC1A 7AU

Dept. of Energy,
Thames House, South Millbank,
London SW1P 4QJ

National Coal Board, Films Officer
Hobart House, Grosvenor Place,
London SW1 7AE

Shell Education Service,
Shell Centre,
London SE1 7NA

Film resource
Cheddar Gorge Caves
(16 mm, colour, free of charge)
from:
The Manager,
Cheddar Caves,
Somerset.

ANSWERS TO EXERCISES

CHAPTER 1: Exercise 1

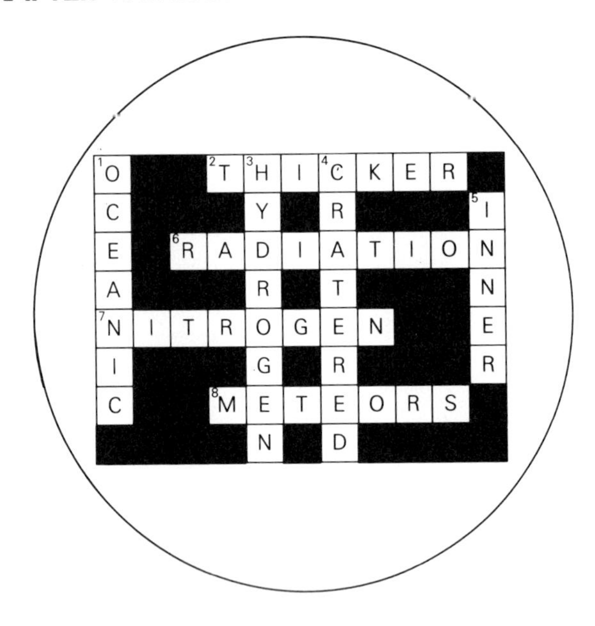

CHAPTER 2: Exercise 1

CHAPTER 3: Exercise 1

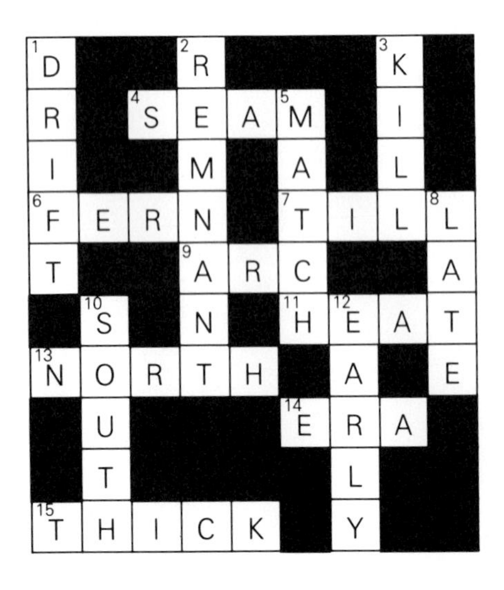

CHAPTER 4: Exercise 1

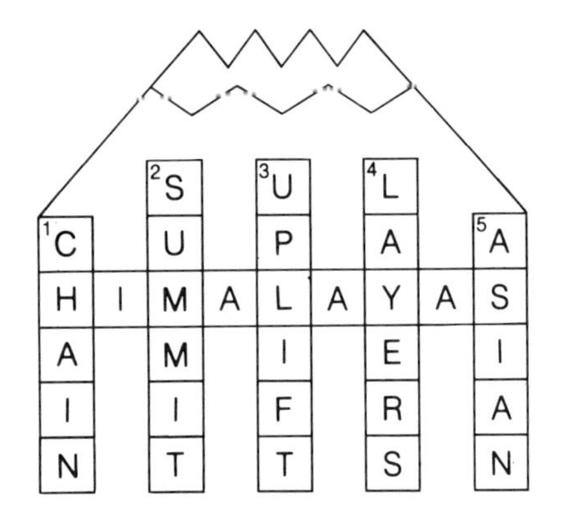

CHAPTER 5: Exercise 1

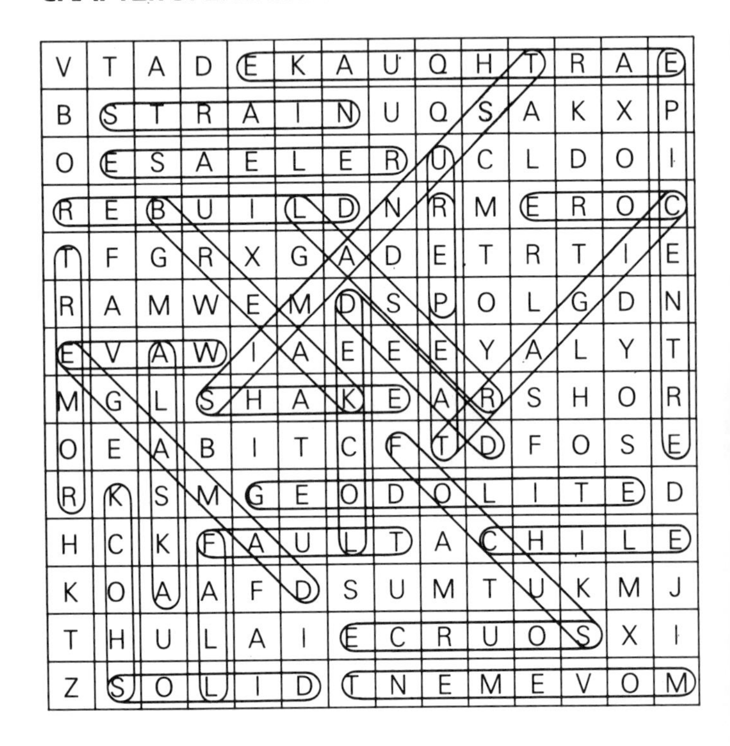

CHAPTER 6: Exercise 1

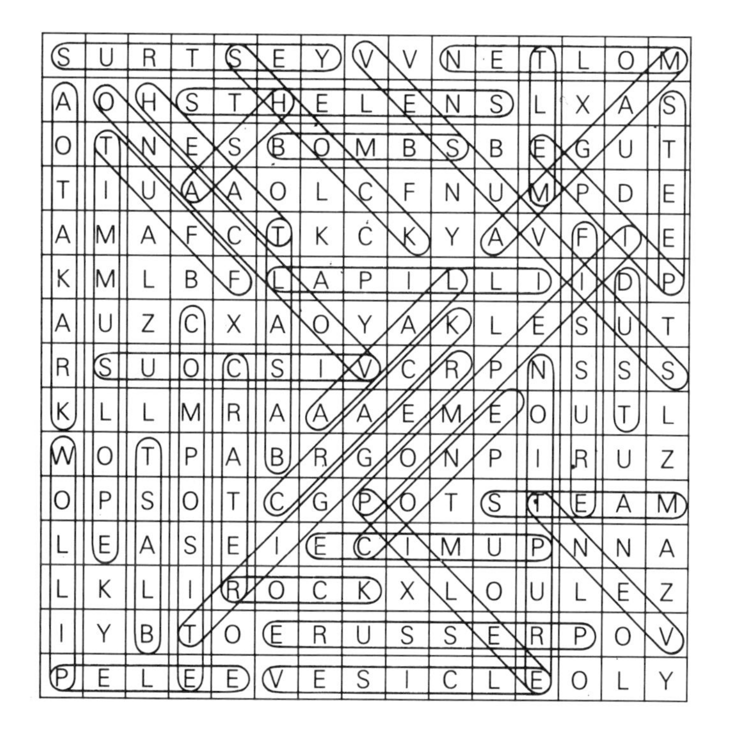

CHAPTER 8: Exercise 1

CHAPTER 10: Exercise 1

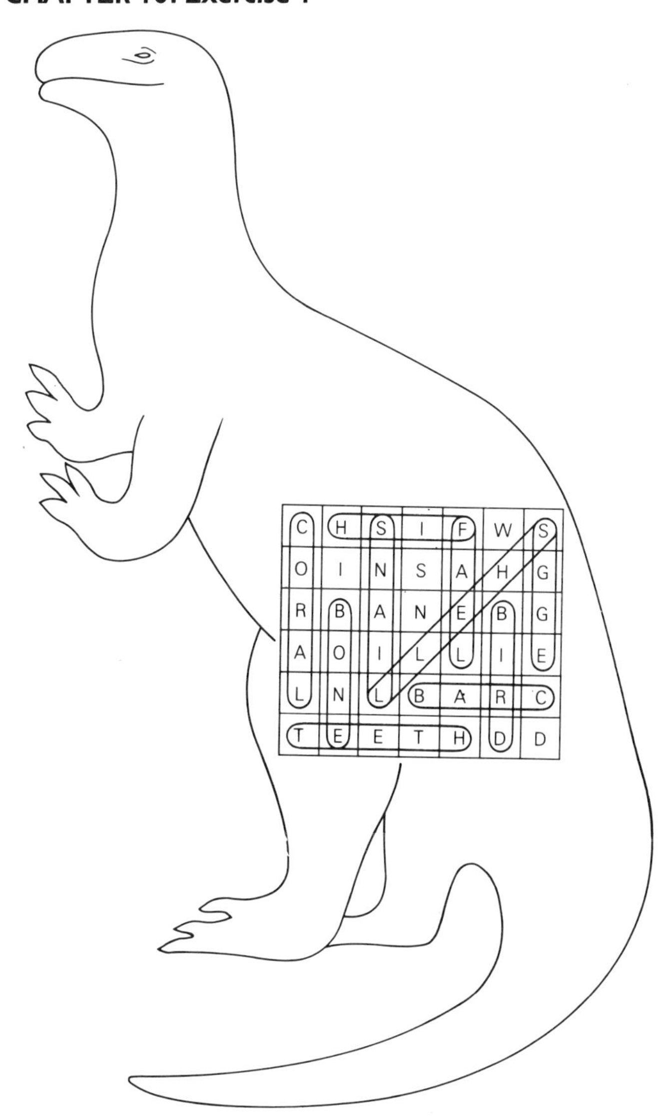

CHAPTER 12: Exercise 1

CHAPTER 13: Exercise 1

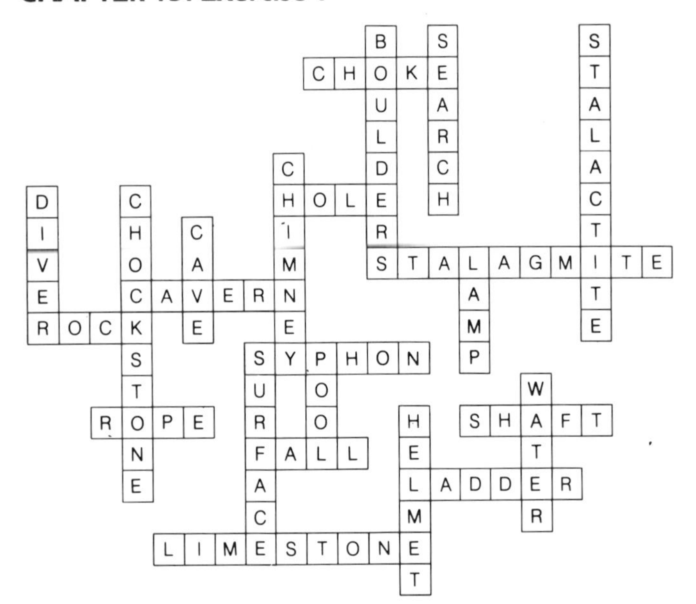

CHAPTER 14: Exercise 1

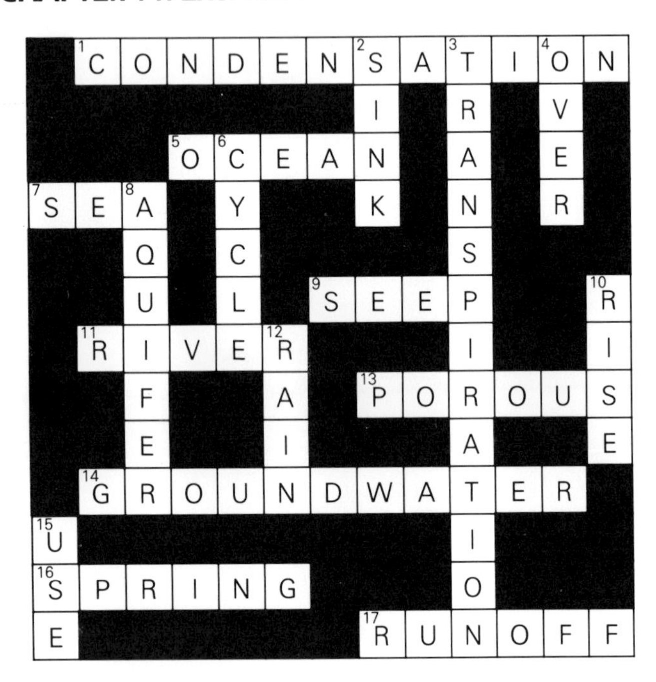

CHAPTER 16: Exercise 1

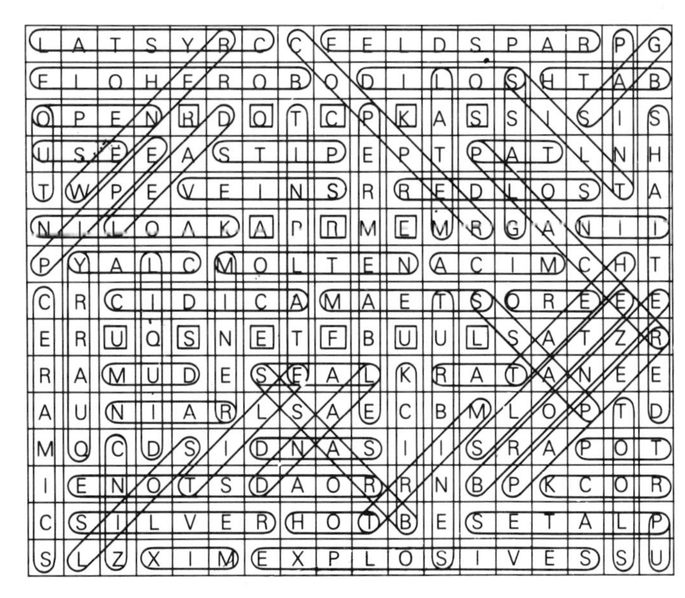

General

The Age of the Earth, The Geological Museum (HMSO)
Elements of Field Geology, Himus and Sweeting (UTP)

Plate tectonics, mountain building and continental drift

Drifting Continents, Shifting Seas, P. Young (Watts)
The Evolving Earth, Sawkins *et al* (Collier Macmillan)
Plate Tectonics, D. C. Heather (Edward Arnold)
The Story of the Earth, Geological Museum (HMSO)
Tectonic Processes, D. Weyman (Allen and Unwin)

Earthquakes and volcanoes

Catastrophe: The Violent Earth, T. Waltham (Macmillan)
Earthquakes, The Geological Museum (HMSO)
Volcanoes, P. Francis (Penguin)

Minerals, rocks and fossils

British Cainozoic Fossils, British Mesozoic Fossils and *British Palaeozoic Fossils* The British Museum (HMSO). Three excellent reference books for fossil collectors.
Elements of Palaeontology, R. M. Black (Cambridge University Press)
Finding Fossils, R. Hamilton and A. Insole (Penguin)
Fossils, D. Dineley (Collins)
The Hamlyn Guide to Minerals, Rocks and Fossils, W. R. Hamilton *et al* (Hamlyn)
Lucy – The Beginnings of Humankind, D. C. Johanson (Paladin)
Rocks, D. Dineley (Collins)

Britain: Past and present

British Regional Geology Guides (HMSO)
Geology and Scenery in England and Wales, A. E. Trueman (Pelican)
Geology and Scenery in Scotland, A. E. Trueman (Pelican)
Landscape Studies, O. Sawyer (Edward Arnold)
On the Rocks (BBC Publications)
Physical Geography in Diagrams, R. B. Bunnett (Longman)

Resources

Earth Resources – A dictionary of terms and concepts, D. Dineley *et al* (Arrow)
Energy: Needs and Resources, P. Odell (Macmillan Education)
Mineral Resources Book, Science in Society (Heinemann)

INDEX

acid rock 56
alluvium 109
ammonite 82
Anatolian fault 25
Andes 23
andesite 56, 115
anticline 21, 22, 135
aquifer 107
arch 120
Archaeopteryx 86
arete 119
artesian basin 107
asteroid 2
atmosphere 2
atoll 95
atoms, structure of 43
augite 44
aureole, metamorphic 61, 124
avalanche 29

bar 121
barytes 44
basalt 5–6, 36, 54, 117
basic rocks 56
bathylith 40
bauxite 90
bedding plane 57
belemnite 81
bivalve 80
blocky lava 33
bomb, volcanic 34
borax 98
boulder clay 119
brachiopod 80
bronze 90

calcite 44
calcium carbonate 101
calcium hydrogencarbonate 102
caldera 36
Caledonian Mountains 115
carbon 43
Carboniferous limestone 102, 116
carbonisation 78
catchment 109
cavern 102
cephalopod 81
chalk 47, 100, 117
Charnia 85
Cheddar Gorge 102
Chesil Beach 121
china clay 125
cirque glacier 118
cleavage 47, 62
clint 102
coal 17, 129
Coal Measures 116
coccolith 100
columnar jointing 54
composite cone 34, 35
conglomerate 58
continental drift 16
convection current 5, 9
coprolite 76
coral 80
coral reef 95
core 5, 29
correlate 86
crater 4, 34
crinoid 83
crust 5
crystal 42, 55
current bedding 58
cyclothem 116

deep sea 95
delta 58, 95, 109
density 5, 46
desalination 98
diamond 3, 43–4
dinosaur 76
diorite 56
Diplodocus 85
distillation 137
dolerite 54
dolomite 44
drainage basin 109
dry valley 102
dyke 40

earthquake 9, 25–31
echinoderm 83
echinoid 83
El Chichon, Mexico 37–8
element 43
epicentre 26
era 4
erosion 4, 6, 63–4
erratics 120
eruption 33
Erzurum, Turkey 25
escarpment 102, 120
evaporite 97
Everest, Nepal 21
evolution 84
extrusive rock 54

facies 114
fault 9, 134
feldspar 44
felsic mineral 56
fjord 121
fissure 12
flint 47, 48
flocculation 70
flood-plain 109
fluorspar 44, 47
focus 26
fold 6, 21
—— mountains 21
foraminifera 96
fossil 17, 60, 76–87
—— fuel 130
fracture, of minerals 47
frost shattering 65

gabbro 56
galaxy 2
galena 44
gastropod 83
geochemical survey 92
geothermal energy 126
geodolite 31
geological time 4
glacier 119
glacials 86, 119
glaciation 119
Glossopteris 17
glowing avalanche 34, 35
gneiss 61, 62
gold 43, 97
Gondwanaland 16
graded bedding 58
granite 42, 66, 124
graphite 43–4
graptolite 84
gravity 2

gravimeter 91
Great Australian Barrier Reef 95
Great Whin Sill 40
grike 102
groundwater 106
Gulf Stream 18
gypsum 44, 97

habit, of minerals 48
hardness, mineral test 44
haematite 44, 48
halides 44
halite 44, 48
Hawaii 29–30, 35
Himalayas 21, 22
hominid 79
horizon (soil) 69
hornblende 44, 47
hornfels 61
hot rock project, Cornwall 127
hot spot 36
humus 69
hydrocarbon 133
hydro-thermal 44

Iceland 12, 39
igneous rock, 42, 54
ignimbrite 34
impermeable rock 106
India 22
intensity (of earthquakes) 26
interglacial 86, 118
intermediate igneous rock 56
intrusion 40
intrusive rocks 54
iridium clay layer 86
island arc 13
isoseismal line 26
isostasy 6, 121

kaolin (china clay) 126
Krakatoa, Java 34–5

L wave 28–9
landslide 29
lapilli 34
Laurasia 16
lava 11–12, 33
—— lake 36
leaching 70
levée 109
Lewisian Gneiss 115
lichen 69
limestone 60, 100–104
Lingula 87
load, of rivers 109
loam 70
longshore drift 121
Lucy 79
Lulworth Cove, Dorset 121
lustre 46

mafic mineral 56
magma 5, 9, 36, 54
Magnesian limestone 60, 101
magnetic field 5
—— record 17
—— stripes 12
—— survey 92
magnetite 44
magnitude, of earthquakes 26–7
malachite 44
mantle 5, 29
marble 61
Marianas trench 13
Mauna Loa, Hawaii 35, 36
meander 109
Mediterranean Sea 17
Mercalli scale 26
Mesosaurus 17
metamorphic rock 54, 61–2
Meteor Crater, Arizona 3
meteorite 2–4
Mendip Hills 101, 126
mica 44, 47
microdiorite 56
microgranite 56
Micraster 84–5, 86
mineral 42
—— vein 44, 124
Millstone Grit 116
Mohorovicic 29
Mohs scale 44
moraine 119
Mt Pelee, Martinique 35
Mt St Helens, USA 36
mudstone 58

natural selection 84
nebula 4
New Red Sandstone 116
nitrogen 2
nodule (metallic) 96
North Sea oil 135
nuée ardente 35, 37
obsidian 47, 54
ocean ridge 11
oil 133
—— trap 134
Old Red Sandstone 115
olivine 44
oolith 101
oolitic limestone 101
ooze 96
opencast mining 92, 131
ore 90
orthoclase feldspar 44
ox-bow lake 108

P wave 28
palaeogeographic map 115
palaeomagnetism 115
Panama isthmus 18, 19
Pangaea 16
panning 91
pavement, limestone 102
period 4
pervious 106
petrification 78
petroleum 133
petrochemicals 136
phenocryst 55
pillow lava 12
placer 91
plastic flow 5
plate tectonics 9
plagioclase feldspar 44
pluton 14, 22
Popayan, Columbia 27
porous rock 106
porphyritic texture 55
pothole 108
Pozzuoli, Italy 39
pumice 34
pyrite 44

quartz 42
quartzite 61
quicklime 104

radiometric dating 87
raised beach 122
recumbent fold 22
red clay 96
Red Sea 18
regional metamorphism 62
relative dating 86
reniform habit 48

rhyolite 56
ria 121
Richter scale 26
rift valley 11, 108
ripple mark 58, 114
river 107
rock cycle 63
rock salt 43, 48
ropy lava 33

S wave 28
salinity 98
San Andreas 14, 30
San Francisco 14, 30
sandstone 58
schist 62
scree 65
sea-floor spreading 13
seam 130
sea-pen 85
seat earth 130
seawater 97
sediment 57
sedimentary rock 54, 57
seismic wave 26, 27–8
seismic prospecting 91, 135
seismometer 26, 27–8
shadow zone 28
shale 58, 61
shelf sea 95
shield area 6
shield volcano 35
silica 56
silicate 44
silicon-oxygen tetrahedra 44
sill 40
slate 61, 124
smoker vent 97
sodium chloride 97
soil conservation 71
—— creep 71
—— erosion 71
—— formation 69
—— profile 69
solar system 2
solution mining 93
sphalerite 44, 91
spit 121
spring 107
——, hot 127
stack 60, 120
stalactite 102
stalagmite 102
star 2
steel 90
strata 113
stratigraphy 113
streak 46
striation 119
subduction 13, 22
submarine canyon 96
sulphates 44
Superposition, Rule of 113
swallow hole 102
syncline 21, 22

tarn 118
Tethys ocean 16, 17, 22
thrust fault 22
till 16, 119
tillite 120
tiltmeter 39
tin 90, 124
tor 66
trace, fossil 76
transform fault 10, 14
trench 13
trilobite 83
tsunamis 29
tuff 34
turbidity current 95
Tyrannosaurus 76

unconformity 113, 135

Variscan mountain range 116
vein 44, 124
vent 33
vesicle 34
Vesuvius, Italy 39
volcano 9, 33

water 106
watershed 109
water table 106
weathering 63
——, of limestone 102
welded-tuff 34
William Smith 86
Wray, North Lancashire 109

zonal dating 86